Meta–Learning Frameworks for Imaging Applications

Ashok Sharma
University of Jammu, India

Sandeep Singh Sengar
Cardiff Metropolitan University, UK

Parveen Singh
Cluster University, Jammu, India

A volume in the Advances in
Computational Intelligence and
Robotics (ACIR) Book Series

Published in the United States of America by
    IGI Global
    Engineering Science Reference (an imprint of IGI Global)
    701 E. Chocolate Avenue
    Hershey PA, USA 17033
    Tel: 717-533-8845
    Fax: 717-533-8661
    E-mail: cust@igi-global.com
    Web site: http://www.igi-global.com

Library of Congress Cataloging-in-Publication Data

Names: Sharma, Ashok, 1977- editor. | Sengar, Sandeep, 1985- editor. |
   Singh, Parveen, 1976- editor.
Title: Meta-learning frameworks for imaging applications / edited by: Ashok
   Sharma, Sandeep Sengar, Parveen Singh.
Description: Hershey PA : Engineering Science Reference, [2023] | Includes
   bibliographical references. | Summary: "The book will bring together
   many experts from the domains of machine learning and imaging
   application to explore the current state of Meta-Learning, its
   application to medical imaging and health informatics, and its future
   directions. This book will give an overview of the Meta- Learning
   framework in Imaging Application"-- Provided by publisher.
Identifiers: LCCN 2023021240 (print) | LCCN 2023021241 (ebook) | ISBN
   9781668476598 (hardcover) | ISBN 9781668476604 (paperback) | ISBN
   9781668476611 (ebook)
Subjects: LCSH: Diagnostic imaging. | Machine learning. | Deep learning
   (Machine learning)
Classification: LCC RC78.7.D53 M48 2023 (print) | LCC RC78.7.D53 (ebook)
   | DDC 616.07/54--dc23/eng/20230607
LC record available at https://lccn.loc.gov/2023021240
LC ebook record available at https://lccn.loc.gov/2023021241

This book is published in the IGI Global book series Advances in Computational Intelligence and
Robotics (ACIR) (ISSN: 2327-0411; eISSN: 2327-042X)

British Cataloguing in Publication Data
A Cataloguing in Publication record for this book is available from the British Library.

For electronic access to this publication, please contact: eresources@igi-global.com.

Advances in Computational Intelligence and Robotics (ACIR) Book Series

ISSN:2327-0411
EISSN:2327-042X

Editor-in-Chief: Ivan Giannoccaro, University of Salento, Italy

MISSION

While intelligence is traditionally a term applied to humans and human cognition, technology has progressed in such a way to allow for the development of intelligent systems able to simulate many human traits. With this new era of simulated and artificial intelligence, much research is needed in order to continue to advance the field and also to evaluate the ethical and societal concerns of the existence of artificial life and machine learning.

The **Advances in Computational Intelligence and Robotics (ACIR) Book Series** encourages scholarly discourse on all topics pertaining to evolutionary computing, artificial life, computational intelligence, machine learning, and robotics. ACIR presents the latest research being conducted on diverse topics in intelligence technologies with the goal of advancing knowledge and applications in this rapidly evolving field.

COVERAGE

- Algorithmic Learning
- Computational Intelligence
- Intelligent Control
- Evolutionary Computing
- Automated Reasoning
- Agent technologies
- Computational Logic
- Heuristics
- Fuzzy Systems
- Natural Language Processing

IGI Global is currently accepting manuscripts for publication within this series. To submit a proposal for a volume in this series, please contact our Acquisition Editors at Acquisitions@igi-global.com or visit: http://www.igi-global.com/publish/.

Titles in this Series

For a list of additional titles in this series, please visit:
http://www.igi-global.com/book-series/advances-computational-intelligence-robotics/73674

Advanced Interdisciplinary Applications of Machine Learning Python Libraries for Data Science
Soly Mathew Biju (University of Wollongong in Dubai, UAE) Ashutosh Mishra (Yonsei University, South Korea) and Manoj Kumar (University of Wollongong in Dubai, UAE)
Engineering Science Reference • copyright 2023 • 304pp • H/C (ISBN: 9781668486962) • US $275.00 (our price)

Recent Developments in Machine and Human Intelligence
S. Suman Rajest (Dhaanish Ahmed College of Engineering, India) Bhopendra Singh (Amity University, Dubai, UAE) Ahmed J. Obaid (University of Kufa, Iraq) R. Regin (SRM Institute of Science and Technology, Ramapuram, India) and Karthikeyan Chinnusamy (Veritas, USA)
Engineering Science Reference • copyright 2023 • 359pp • H/C (ISBN: 9781668491898) • US $270.00 (our price)

Advances in Artificial and Human Intelligence in the Modern Era
S. Suman Rajest (Dhaanish Ahmed College of Engineering, India) Bhopendra Singh (Amity University, Dubai, UAE) Ahmed J. Obaid (University of Kufa, Iraq) R. Regin (SRM Institute of Science and Technology, Ramapuram, India) and Karthikeyan Chinnusamy (Veritas, USA)
Engineering Science Reference • copyright 2023 • 409pp • H/C (ISBN: 9798369313015) • US $300.00 (our price)

Handbook of Research on Advancements in AI and IoT Convergence Technologies
Jingyuan Zhao (University of Toronto, Canada) V. Vinoth Kumar (Jain University, India) Rajesh Natarajan (University of Applied Science and Technology, Shinas, Oman) and T.R. Mahesh (Jain University, India)
Engineering Science Reference • copyright 2023 • 372pp • H/C (ISBN: 9781668469712) • US $380.00 (our price)

Scalable and Distributed Machine Learning and Deep Learning Patterns

For an entire list of titles in this series, please visit:
[go here and find specific BS URL: http://www.igi-global.com/book-series/]

701 East Chocolate Avenue, Hershey, PA 17033, USA
Tel: 717-533-8845 x100 • Fax: 717-533-8661
E-Mail: cust@igi-global.com • www.igi-global.com

Table of Contents

Preface..xiv

Chapter 1
Optimization Approaches in Meta-Learning Models 1
    Nidhi Arora, Lovely Professional University, India
    Ashok Sharma, University of Jammu, India
    Dinesh Kumar, Maharaja Ranjit Singh Punjab Technical University,
        India

Chapter 2
A Review on Image Super-Resolution Using GAN..12
    Ajay Sharma, VIT Bhopal University, India
    Bhavana Shrivastava, Maulana Azad National Institute of Technology,
        India
    Swati Gautam, Maulana Azad National Institute of Technology, India

Chapter 3
Optimizing Hyper Meta Learning Models: An Epic 32
    G. Devika, Government Engineering College, Krishnarajapete, India
    Asha gowda Karegowda, Siddaganga Institute of Technology, India

Chapter 4
A Study on Evaluation and Analysis of Edge Detection Operators....................65
    Pinaki Pratim Acharjya, Haldia Institute of Technology, India
    Subhankar Joardar, Haldia Institute of Technology, India
    Santanu Koley, Haldia Institute of Technology, India
    Subhabrata Barman, Haldia Institute of Technology, India

Chapter 5

Augmented Reality and Its Significance in Healthcare Systems 103
    Ashish Tripathi, School of Computing Science and Engineering,
        Galgotias University, Greater Noida, India
    Nikita Chauhan, G.L. Bajaj Institute of Technology and Management,
        Greater Noida, India
    Arjun Choudhary, Sardar Patel University of Police, Security, and
        Criminal Justice, Jodhpur, India
    Rajnesh Singh, School of Computing Science and Engineering,
        Galgotias University, Greater Noida, India

Chapter 6

Automated Diagnosis of Eye Problems Using Deep Learning Techniques on
Retinal Fundus Images 119
    N. Sasikaladevi, SASTRA University (Deemed), India
    S. Pradeepa, SASTRA University (Deemed), India
    K. Malvika, SASTRA University (Deemed), India

Chapter 7

Comparative Analysis and Automated Eight-Level Skin Cancer Staging
Diagnosis in Dermoscopic Images Using Deep Learning 134
    Auxilia Osvin Nancy V., Department of Computer science and
        Engineering, College of Engineering and Technology, SRM Institute
        of Science and Technology, Vadapalani Campus, Chennai, India
    P. Prabhavathy, Department of Computer science and Engineering,
        College of Engineering and Technology, SRM Institute of Science
        and Technology, Vadapalani Campus, Chennai, India
    Meenakshi S. Arya, Department of Transportation, Iowa State
        University, USA
    B. Shamreen Ahamed, Deparment of Computer science and
        Engineering, College of Engineering and Technology, SRM Institute
        of Science and Technology, Vadapalani Campus, Chennai, India

Chapter 8

Critical Review Analysis on Deep Learning-Based Segmentation Techniques
for Water-Body Extraction 153
    Swati Gautam, Maulana Azad National Institute of Technology, India
    Ajay Sharma, VIT Bhopal University, India
    Bhavana Prakash Shrivastava, Maulana Azad National Institute of
        Technology, India

Chapter 9
Hybrid Approaches for Plant Disease Recognition: A Comprehensive Review169
S. Hemalatha, School of Computer Science Engineering and
Information Systems, Vellore Institute of Technology, India
Athira P. Shaji, School of Computer Science Engineering and
Information Systems, Vellore Institute of Technology, India

Chapter 10
Redefining Management With Advent of Artificial Intelligence in the Current
Business World 186
Rohit Bhagat, University of Jammu, India
Vinay Chauhan, University of Jammu, India

Compilation of References 198

Related References 222

About the Contributors 247

Index 252

Detailed Table of Contents

Preface.. xiv

Chapter 1

Optimization Approaches in Meta-Learning Models 1

    Nidhi Arora, Lovely Professional University, India
    Ashok Sharma, University of Jammu, India
    Dinesh Kumar, Maharaja Ranjit Singh Punjab Technical University,
        India

This book chapter provides a comprehensive overview of optimization approaches in meta-learning, focusing on techniques and their applications. Meta-learning is a subfield of machine learning that emphasizes acquiring knowledge from previous tasks and applying the same to new tasks in order to develop the models with improved learning process. Optimization plays a crucial role in meta-learning models by enabling the effective acquisition and utilization of knowledge across tasks. This chapter provides an overview of various optimization approaches employed in meta-learning models which entail changing the model's input parameters or learning algorithms to facilitate effective learning across various tasks or domains. The methods tackle the problem of effective learning without compromising with accuracy and precision in performance focusing on the benefits of meta-learning frameworks in practical situations which may be considered as the real-world applications of these approaches.

Chapter 2

A Review on Image Super-Resolution Using GAN... 12

    Ajay Sharma, VIT Bhopal University, India
    Bhavana Shrivastava, Maulana Azad National Institute of Technology,
        India
    Swati Gautam, Maulana Azad National Institute of Technology, India

This study focuses on the utilization of generative adversarial networks (GANs) for generating high-resolution facial images from low-resolution inputs, which is vital for computer vision applications. Facial images present a complex structure, posing challenges for obtaining high-quality results using traditional super-resolution methods. However, recent advancements in deep learning, particularly GANs, have shown promising outcomes in this area. In this work, the authors conduct a comprehensive analysis of state-of-the-art GAN-based techniques for realistic high-resolution face image generation. They discuss the principles of image degradation, the learning process of GANs, and the challenges associated with these methods. By offering insights into the current state and future research directions, they aim to familiarize readers with the context and significance of GAN-based face image generation. This work highlights the importance of GANs in improving facial image quality and their relevance to advancing computer vision applications such as face verification and recognition.

Chapter 3

Optimizing Hyper Meta Learning Models: An Epic 32
G. Devika, Government Engineering College, Krishnarajapete, India
Asha gowda Karegowda, Siddaganga Institute of Technology, India

Optimizing hyper meta learning models is a critical task in the field of machine learning, as it can improve the performance, efficiency, and scalability of these models. In this chapter, the authors present an epic overview of the process of optimizing hyper meta learning models. They discuss the key steps involved in this process, including task selection, model architecture selection, hyperparameter optimization, model training, model evaluation, and deployment. They also explore the benefits of hyper meta learning models and their potential future applications in various fields. Finally, they highlight the challenges and limitations of hyper meta learning models and suggest future research directions to overcome these challenges and improve the effectiveness of these models.

Chapter 4

A Study on Evaluation and Analysis of Edge Detection Operators....................65
Pinaki Pratim Acharjya, Haldia Institute of Technology, India
Subhankar Joardar, Haldia Institute of Technology, India
Santanu Koley, Haldia Institute of Technology, India
Subhabrata Barman, Haldia Institute of Technology, India

One of the key stages in both image processing and computer vision is edge detection. For analysis and measurement of several fundamental attributes of an object or set of objects in an image, such as area, perimeter, and form, correct identification of the edges of the objects in the image is crucial. The edge detection operators employed in image processing must therefore be thoroughly understood. In this chapter, fundamental theories and comparative assessments of several edge detection operators are discussed along with a proposed improved contour detection scheme for better performance measurement. The technique has been used to process a number of digital photos, and improved performance in terms of contour detection has been attained.

Chapter 5

Augmented Reality and Its Significance in Healthcare Systems 103
    Ashish Tripathi, School of Computing Science and Engineering,
        Galgotias University, Greater Noida, India
    Nikita Chauhan, G.L. Bajaj Institute of Technology and Management,
        Greater Noida, India
    Arjun Choudhary, Sardar Patel University of Police, Security, and
        Criminal Justice, Jodhpur, India
    Rajnesh Singh, School of Computing Science and Engineering,
        Galgotias University, Greater Noida, India

Augmented reality and virtual reality are terms often used together and even interchangeably sometimes without knowing their actual meaning. Augmented reality (AR) enhances the real world by mixing and overlapping digital objects with the real world whereas virtual reality (VR) is a completely different world created in a virtual space. VR can be experienced with wearables; but AR needs a device as simple as a phone and it's also wearable. In this chapter, AR is discussed and explored in a detailed manner. With its rapid evolving time, AR will be more common than it is now. It already is a part of everyone's life with the help of applications like Google Lens and Snapchat. AR has been experimented on for a while and the first spine surgery on a patient has been performed by John Hopkins neurosurgeon on June 8, 2020, using AR headsets. In this chapter, the types of extended reality (XR) and their differences. AR technology used in the study of anatomy, medical surgeries, pharma study, MedTech, and case studies of AR implementation in the field of medical surgeries is discussed.

Chapter 6

Automated Diagnosis of Eye Problems Using Deep Learning Techniques on Retinal Fundus Images...119
    N. Sasikaladevi, SASTRA University (Deemed), India
    S. Pradeepa, SASTRA University (Deemed), India
    K. Malvika, SASTRA University (Deemed), India

Automated diagnosis of eye diseases using deep learning techniques on retinal fundus images has become an active area of research in recent years. The suggested method divides retinal images into various disease categories by extracting relevant data using convolutional neural network (CNN) architecture. The dataset used in this study consists of retinal images taken from patients with various eye conditions, such as age-related macular degeneration, glaucoma, and diabetic retinopathy. The aim of this study is to investigate the potential of deep learning algorithms in detecting and classifying various retinal diseases from fundus images. The suggested approach may make early eye disease diagnosis and treatment easier, reducing the risk of vision loss and enhancing patient quality of life. The DenseNet-201 model is tested and achieved an accuracy rate of 80.06%, and the findings are extremely encouraging.

Chapter 7

Comparative Analysis and Automated Eight-Level Skin Cancer Staging
Diagnosis in Dermoscopic Images Using Deep Learning 134

   Auxilia Osvin Nancy V., Department of Computer science and
      Engineering, College of Engineering and Technology, SRM Institute
      of Science and Technology, Vadapalani Campus, Chennai, India
   P. Prabhavathy, Department of Computer science and Engineering,
      College of Engineering and Technology, SRM Institute of Science
      and Technology, Vadapalani Campus, Chennai, India
   Meenakshi S. Arya, Department of Transportation, Iowa State
      University, USA
   B. Shamreen Ahamed, Deparment of Computer science and
      Engineering, College of Engineering and Technology, SRM Institute
      of Science and Technology, Vadapalani Campus, Chennai, India

The challenge in the predictions of skin lesions is due to the noise and contrast. The manual dermoscopy imaging procedure results in the wrong prediction. A deep learning model assists in detection and classification. The structure in the proposed handles CNN architecture with the stack of separate layers that use a differential function to transform an input volume into an output volume. For image recognition and classification, CNN is specifically powerful. The model was trained using labeled data with the appropriate class. CNN studies the relationship between input features and class labels. For model building, use Keras for front-end development and Tensor Flow for back-end development. The first step is to pre-process the ISIC2019 dataset, splitting it into 80% training data and 20% test data. After the training and test splits are complete, the dataset has been given to the CNN model for evaluation, and the accuracy on each lesion class was calculated using performance metrics. The comparative analysis has been done on pretrained models like VGG19, VGG16, and MobileNet.

Chapter 8

Critical Review Analysis on Deep Learning-Based Segmentation Techniques
for Water-Body Extraction 153

   Swati Gautam, Maulana Azad National Institute of Technology, India
   Ajay Sharma, VIT Bhopal University, India
   Bhavana Prakash Shrivastava, Maulana Azad National Institute of
      Technology, India

The rapid advancement in the applications of remote sensing imagery had attracted considerable attention from researchers for digital image analysis. Researchers had performed the surveying and delineation of water bodies with excellent efforts and algorithms in the past, but they faced many challenges due to the varying characteristics of water such as its shape, size, and flow. Traditional methods employed for water body segmentation posed certain limitations in terms of accuracy, reliability, and robustness. Rapid growth in the automation category allowed researchers to incorporate deep learning models into the segmentation analysis. Deep learning segmentation models for water body feature extraction have shown promising results based on accuracy and precision. This chapter presents a brief review on the deep learning models used for water-body extraction with their merits over the traditional approaches. It also discusses existing results with challenges faced and future scope.

Chapter 9

Hybrid Approaches for Plant Disease Recognition: A Comprehensive Review169
S. Hemalatha, School of Computer Science Engineering and
Information Systems, Vellore Institute of Technology, India
Athira P. Shaji, School of Computer Science Engineering and
Information Systems, Vellore Institute of Technology, India

Plant diseases pose a significant threat to agriculture, leading to yield and quality losses. Traditional manual methods for disease identification are time-consuming and often yield inaccurate results. Automated systems leveraging image processing and machine learning techniques have emerged to improve accuracy and efficiency. Integrating these approaches allows image preprocessing and feature extraction to be combined with machine learning algorithms for pattern recognition and classification. Deep learning, particularly convolutional neural networks (CNNs), has revolutionized computer vision tasks, enabling hierarchical feature extraction. Hybrid methods offer advantages such as improved accuracy, faster identification, cost reduction, and increased agricultural productivity. This survey explores the significance and potential of hybrid approaches in plant disease identification, addressing the growing need for early detection and management in agriculture.

Chapter 10

Redefining Management With Advent of Artificial Intelligence in the Current Business World 186
Rohit Bhagat, University of Jammu, India
Vinay Chauhan, University of Jammu, India

The emergence of artificial intelligence has been a start of new era in the age of technology. The business world mostly relies on satisfying customer needs, with the introduction of artificial intelligence the task of marketer to satisfy customer has been made much simpler. The use of artificial intelligence has added a lot of value to the customers while purchasing, which has added to an overall increase in customer experience. Decision making is one of the most difficult tasks from the customer's point of view; when it comes to choosing a product, the use of artificial intelligence has been very fruitful in helping the customer in making decisions. The chapter tries to show how the use of artificial intelligence has changed the way of doing business. The chapter concludes with the future scope of artificial intelligence in the field of management and its application.

Compilation of References 198

Related References 222

About the Contributors 247

Index 252

Preface

Meta-learning, a concept often described as "learning to learn," has emerged as a compelling paradigm in the world of artificial intelligence and machine learning. Instead of starting from scratch for every new machine learning task, meta-learning offers a systematic and efficient way to adapt and learn from previous experiences. This approach not only accelerates the development of machine learning systems but also allows us to replace handcrafted algorithms with data-driven solutions.

In recent years, the field of meta-learning has gained significant traction and has been widely embraced by researchers and practitioners alike. Its applications span various domains, including computer vision, healthcare, and beyond. This edited reference book, *Meta-Learning Frameworks for Imaging Applications*, authored and curated by Ashok Sharma, Sandeep Sengar, and Parveen Singh, is a comprehensive exploration of meta-learning with a specific focus on imaging applications.

Deep neural networks have revolutionized machine learning, but they come with their own set of challenges, such as high data requirements, computationally expensive training, and limited task transferability. The meta-learning framework is poised to tackle these issues head-on.

In this book, readers will find an in-depth overview of the meta-learning framework, its principles, and its relevance in the context of imaging applications. The book chapters cover a spectrum of meta-learning approaches, including model-agnostic learning, memory augmentation, prototype networks, and learning to optimize, providing readers with a holistic understanding of the subject.

One of the distinctive features of this book is its emphasis on the practical aspects of meta-learning. It delves into experimental techniques and evaluation methodologies critical for implementing and assessing meta-learning frameworks effectively. The goal is to equip readers with the knowledge and tools needed to harness the power of meta-learning in their own research and applications.

Furthermore, this book assembles a diverse group of experts from the fields of machine learning and imaging applications. They share their insights, experiences, and cutting-edge research to provide a comprehensive snapshot of the current state of meta-learning in the realm of medical imaging and health informatics. Moreover, the book explores the future directions and potential innovations that lie ahead in this exciting field.

This book is designed to serve as a valuable resource for a wide audience, including machine learning researchers, biomedical engineers, medical practitioners, and graduate students. Whether you are new to the concept of meta-learning or looking to deepen your knowledge and apply it to medical imaging and health informatics, this book offers a comprehensive introduction and a rich source of information.

Chapter 1

In this introductory chapter, we delve into the fundamental concepts of meta-learning and its emphasis on knowledge acquisition from previous tasks to enhance the learning process. Specifically, we explore the critical role of optimization techniques in meta-learning models. This chapter provides an extensive overview of various optimization approaches employed in meta-learning, shedding light on how they modify the input parameters and learning algorithms to facilitate effective learning across diverse tasks and domains. Moreover, we highlight the practical implications of these approaches, showcasing their real-world applications and their potential to enhance learning without compromising accuracy and precision.

Chapter 2

In the second chapter, we focus on the exciting realm of generative adversarial networks (GANs) and their application in generating high-resolution facial images from low-resolution inputs. High-quality face image generation is vital for computer vision applications, but traditional methods face challenges with the complex structure of facial images. This chapter offers a comprehensive analysis of state-of-the-art GAN-based techniques for realistic high-resolution face image generation. We explore the principles behind GANs, the learning process involved, and the unique challenges posed by this domain. By providing insights into the current state of GAN-based face image generation, we aim to highlight their significance in advancing computer vision applications, particularly in face verification and recognition.

Chapter 3

The third chapter presents a detailed exploration of the optimization of hyper meta learning models—a critical task in the field of machine learning. We take readers through the essential steps of optimizing these models, from task selection and model architecture to hyperparameter optimization, model training, evaluation, and deployment. We delve into the potential benefits of hyper meta learning models and their future applications across various domains. Moreover, we candidly address the challenges and limitations associated with these models, offering insights into potential research directions to improve their effectiveness.

Chapter 4

Chapter 4 delves into the crucial domain of edge detection, a fundamental stage in image processing and computer vision. Correctly identifying object edges is essential for various image analysis tasks. This chapter provides a comprehensive overview of different edge detection operators, discussing their fundamental theories and offering comparative assessments. Furthermore, it introduces a proposed improved contour detection scheme designed to enhance performance measurement in digital photos. Readers will gain a deeper understanding of the nuances in edge detection techniques and how they can be improved for practical applications.

Chapter 5

Chapter 5 takes readers on an immersive journey into the world of Augmented Reality (AR). It explores the distinctions between AR and Virtual Reality (VR) and delves into the practical applications of AR, which are rapidly evolving and becoming increasingly prevalent in our daily lives. The chapter discusses the types of Extended Reality (XR) and their differences, with a focus on AR's use in anatomy studies, medical surgeries, pharmaceutical research, and MedTech. Readers will also find enlightening case studies showcasing AR implementations in the field of medical surgeries.

Chapter 6

Chapter 6 focuses on the automated diagnosis of eye diseases using deep learning techniques applied to retinal fundus images. The chapter outlines a comprehensive methodology for categorizing various retinal diseases, such as age-related macular degeneration, glaucoma, and diabetic retinopathy. The study investigates the potential of deep learning algorithms in early detection and classification of these diseases from fundus images, with encouraging results. Specifically, the DenseNet-201 model achieves an accuracy rate of 80.06%. This chapter underscores the significance of deep learning in enhancing patient quality of life by facilitating early disease diagnosis and treatment.

Chapter 7

Chapter 7 dives into the realm of skin lesion detection, a challenging task due to noise and contrast variations. The chapter explores the application of deep learning models, specifically Convolutional Neural Networks (CNNs), for accurate detection and classification of skin lesions. It details the model-building process, from data preprocessing to training and evaluation, and offers a comparative analysis of pretrained models like VGG19, VGG16, and MobileNet. This chapter sheds light on how CNNs can revolutionize the field of dermatology by assisting in early skin lesion detection.

Chapter 8

Chapter 8 explores the application of deep learning models in the segmentation and extraction of water bodies from remote sensing imagery. Traditional methods for water body segmentation face challenges in accuracy and robustness. This chapter reviews the merits of deep learning models over traditional approaches, highlighting their potential for accurate water-body feature extraction. The chapter also discusses existing results, challenges faced, and future research directions in this domain.

Chapter 9

In Chapter 9, we address the pressing issue of plant diseases and their impact on agriculture. Traditional manual methods for disease identification are inefficient and often inaccurate. This chapter explores the significance of hybrid approaches that combine image processing and machine learning techniques, particularly Convolutional Neural Networks (CNNs), to improve the accuracy and efficiency of plant disease identification. By integrating image preprocessing, feature extraction, and machine learning algorithms, these hybrid methods offer valuable advantages, including early detection and enhanced agricultural productivity.

Chapter 10

The final chapter delves into the transformative role of artificial intelligence in the world of marketing and management. It discusses how AI has revolutionized customer satisfaction, decision-making, and the overall business landscape. The chapter explores the value AI brings to customers, simplifying their decision-making process. Additionally, it provides insights into the future scope of AI in management and its potential applications. This chapter offers a compelling perspective on the evolving intersection of AI and business, highlighting its transformative potential.

In summary, "Meta-Learning Frameworks for Imaging Applications" offers a unique perspective on meta-learning, its theoretical foundations, practical applications, and its transformative potential in the world of artificial intelligence and healthcare. We hope that readers find this book both informative and inspiring as they embark on their own journey into the fascinating realm of meta-learning.

Ashok Sharma
University of Jammu, India

Sandeep Singh Sengar
Cardiff Metropolitan University, United Kingdom

Parveen Singh
Cluster University Jammu, India

Chapter 1
Optimization Approaches in Meta–Learning Models

Nidhi Arora
Lovely Professional University, India

Ashok Sharma
 https://orcid.org/0000-0003-2553-4263
University of Jammu, India

Dinesh Kumar
Maharaja Ranjit Singh Punjab Technical University, India

ABSTRACT

This book chapter provides a comprehensive overview of optimization approaches in meta-learning, focusing on techniques and their applications. Meta-learning is a subfield of machine learning that emphasizes acquiring knowledge from previous tasks and applying the same to new tasks in order to develop the models with improved learning process. Optimization plays a crucial role in meta-learning models by enabling the effective acquisition and utilization of knowledge across tasks. This chapter provides an overview of various optimization approaches employed in meta-learning models which entail changing the model's input parameters or learning algorithms to facilitate effective learning across various tasks or domains. The methods tackle the problem of effective learning without compromising with accuracy and precision in performance focusing on the benefits of meta-learning frameworks in practical situations which may be considered as the real-world applications of these approaches.

DOI: 10.4018/978-1-6684-7659-8.ch001

1.1 Introduction to Meta-Learning

Meta-learning is the field of machine learning which refers to "learning to learn'', and involves understanding algorithms and metadata. Metadata is data that describes other data. Traditional machine learning algorithms focus on training a model which requires a large dataset. The difficulties that traditional machine learning techniques have when dealing with sparse data and the requirement for ongoing learning are the driving forces behind meta-learning. Gathering significant amounts of labelled data for every new activity or area can often be time-consuming, expensive, or even impractical in real-world situations. This problem is addressed by meta-learning, which uses prior information from related tasks or domains to speed up learning on new, unknown tasks. Meta-learning aims to find the best-performing algorithm and parameters of the algorithm, optimizing the number of experiments and resulting in better predictions in a lesser time (Khan et al., 2020). The algorithms applied in meta learning can adjust optimization and behave to be good at learning with just a few examples. The goal is to make it easy to achieve artificial general intelligence and move artificial intelligence closer to emulating how humans learn and solve problems. Meta-learning acts as a two-level learning process. The model accumulates general knowledge and creates a collection of parameter initializations or representations that are applicable to various tasks at the first level of learning, which entails learning from a set of fundamental tasks or domains. The second level involves modifying the model parameters or changing the representations to swiftly adapt to new tasks or domains, frequently with little data.

1.2 Key Concepts and Approaches

Meta-learning has a variety of approaches, such as metric-based, model-based, and optimization-based methods (Huisman et al., 2021) which helps to find the best-performing algorithm.

1) Meta-learning based on metrics entails the acquisition of a distance measure that may be applied to the comparison of instances from various tasks. The objective is to develop a statistic that the model can use to generalise to new tasks and accelerate learning.

2) Model-based meta-learning: This strategy involves studying a model that may be applied to produce new models for various tasks. The objective is to develop a model that can be generalised to new activities and produce models that excel at those tasks.

3) Optimization-based meta-learning: This method uses optimisation adjustments to learn a model that can be applied to create new models for various tasks. The aim is to learn a model that can be applied to new situations.

In meta-learning models, optimization is essential since it makes it possible to effectively acquire and apply information across tasks. This chapter gives a general overview of the several optimization techniques used in meta-learning models, including gradient-based methods, model-agnostic methods, memory-augmented architectures, and evolutionary procedures.

1.3 Gradient-Based Meta-Learning

1.3.1 Gradient Descent Methods

In the meta-training phase of gradient-based meta-learning, model parameters are updated using gradient descent methods. By iteratively changing the model's parameters in the direction of the steepest fall, gradient descent methods seek to minimise a loss function (Finn & Levine, 2017). These techniques are utilised in the context of meta-learning to identify the best initialization or update rules that allow quick task adaption.

1.3.2 Stochastic Gradient Descent (SGD)

Stochastic gradient descent (Li et al., 2017) is a popular optimisation approach which can handle large datasets, making it suitable for meta-learning scenarios where less data is available for every task. Using a randomly selected subset (mini-batch) of the training data, it calculates the gradients of the loss function with respect to the model parameters. The model parameters are then updated using these gradients.

1.3.3 Adam Optimisation

The optimisation algorithm Adam (Adaptive Moment Estimation) combines the benefits of RMSprop and Adagrad. It keeps track of prior gradients and their squared values and uses those averages, which exhibit exponential decay, to calculate adaptive learning rates for each parameter. Adam is frequently employed in gradient-based meta-learning algorithms and has shown successful in a range of machine learning tasks.

1.3.4 Meta-Gradient Descent

is a crucial element in many meta-learning methods (Schulman et al., 2017). When updating the meta-parameters, it requires computing gradients of the loss function with respect to the original or updated model parameters. The initialization or update rules that the model employs for quick adaption are controlled by the meta-parameters. The model can learn to adapt fast and successfully to new tasks by optimising the meta-parameters through meta-gradient descent.

1.3.5 Practical Considerations

While applying gradient-based meta-learning, several practical considerations should be taken into account (Khodak et al., 2019)

1) Task distribution: The choice of tasks during meta-training should be representative of the target tasks the model will encounter. It is important to ensure diversity in task difficulty, domain, and other relevant factors to encourage robust generalization.
2) Overfitting: Due to the limited number of examples available for each task, these models can prone to overfitting. Regularization techniques, such as weight decay or dropout, can help mitigate overfitting and therefore, help in improving generalization.
3) Meta-learning algorithms frequently use a variety of hyperparameters, including learning rates, regularisation coefficients, and network designs. To achieve good performance, hyperparameters must be tuned properly using methods like grid search or Bayesian optimisation.
4) Efficiency of computation: Gradient-based meta-learning frequently entails several rounds of the inner and outer loops, which can be computationally costly. Efficiency can be increased by employing methods like unrolling processing graphs, reusing intermediate results, or parallel computing.
5) Transferability: A key factor in meta-learning is ensuring that the acquired knowledge or representations may be applied to other tasks. More transferrable models can be created using methods like multi-task learning and unsupervised pre-training.

1.4 Model-Agnostic Meta-Learning (MAML)

Model-Agnostic Meta-Learning (MAML) is a well-known meta-learning optimization technique (Finn et al.,2017) that that may be applied to any algorithm that uses gradient descent to learn. With a few gradient updates, MAML intends to develop a model parameter initialization that can be readily adjusted to new tasks or domains. In order to facilitate quick learning and adaptability, it focuses on identifying a collection of model parameters that can generalise effectively across a variety of tasks (Antoniou et al., 2019).

Consider a classification example where the initial weights are random and the loss is minimised via gradient descent. Gradient descent uses a number of gradient descent steps to discover optimal weights that will result in the least amount of loss in order to approach convergence.

In MAML, these ideal weights are discovered by learning from the distribution of comparable tasks so that starting can be done with optimal weights that will take fewer gradient steps to reach convergence rather than starting with randomly initialised weights.

1.4.1 MAML Algorithm

Meta-training and meta-testing are the two steps that the MAML algorithm takes. (Ye and Chao, 2022)

1) Meta-training: For each meta-training iteration, a batch of tasks is sampled from the task distribution. For each task, the model parameters are cloned and updated using a few gradient steps on a small set of labeled examples (support set). The gradients obtained from the support set are used to update the cloned model parameters. The updated model parameters are then used to evaluate the model's performance on a separate set of labeled examples (query set). The loss on the query set is backpropagated through the entire process to update the meta-parameters (Ravi & Larochelle, 2017).

2) Meta-testing: During meta-testing, new tasks are encountered that were not seen during meta-training. The model parameters are initialized with the learned meta-parameters. The model is fine-tuned or adapted to the new task using a few gradient steps on the support set. The adapted model is then evaluated on the query set to make predictions or perform the desired task.

The MAML algorithm can be summarized as follows:

i.   Initialize the model parameters randomly.
ii.   Sample a batch of tasks from the task distribution.
iii. For each task, compute the gradients of the loss function with respect to the model parameters using a small amount of task-specific data.
iv.   Update the model parameters using the computed gradients.
v.   Repeat steps ii-iv for multiple iterations.

Figure 1. The model-agnostic meta-learning (MAML) algorithm optimizes the parameters θ that can quickly adapt to new tasks
(Finn et al., 2017)

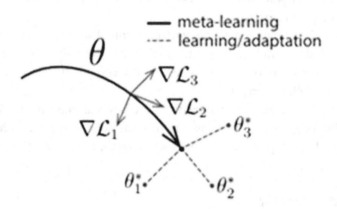

vi. By repeatedly sampling tasks and adapting the model on these tasks, MAML aims to learn a set of initial parameters that can generalize well to new tasks and require only a few gradient steps to adapt. This enables fast adaptation and reduces the need for large amounts of task-specific data.

1.4.2 Quick Adaptation and Applications

Fast adaptability or fine-tuning is a crucial component of MAML. For each new task encountered during meta-testing, it entails updating the model parameters using a limited number of gradient steps on the support set of labelled instances. The goal is to quickly adjust the model using learned meta-parameters as a good initialization to the unique properties of the new task. The model can be improved so that it performs better on the query set of instances for the specified task and generalises well.

MAML has been effectively used in a variety of tasks and domains, such as computer vision, natural language processing, and robotics. Several prominent MAML uses and modifications include:

1) Models that can quickly adapt to new image classification tasks with few labelled examples per class have been learned using MAML.
2) MAML and reinforcement learning have been integrated to allow agents to quickly adapt to new settings or activities without the need for substantial training.

3)   MAML has been expanded to include meta-imitation learning, in which models are taught to copy or imitate expert presentations on brand-new problems with scant data.

4)   In order to increase the model's capacity for adaptability and generalisation, higher-order MAML integrates second-order gradients or curvature information.

1.4.3 Limitations and Challenges

Like any meta-learning system, MAML has its drawbacks and difficulties.

1)   Sensitivity to hyperparameters: The choice of hyperparameters, such as the number of gradient steps or the learning rate during fine-tuning, might affect MAML performance. The performance must be carefully tuned for optimum results.

2)   MAML makes the assumption that the tasks that are encountered during meta-testing are distributed similarly to those in the meta-training set. To tasks that greatly deviate from the training distribution, it might not generalise as effectively.

3)   Sample inefficiency: MAML needs enough gradient steps during meta-training and fine-tuning to be able to adapt to new tasks effectively. For effective adaptation, this can be computationally expensive and may need more data.

4)   Scalability: Scaling MAML to larger models or complex tasks can be challenging due to the computational and memory requirements involved in calculating and storing gradients for multiple tasks.

1.5 Memory-Augmented Meta-Learning

Memory-augmented neural networks are used in Memory-Augmented Meta-Learning, a sort of meta-learning, to quickly assimilate new data and use this data to create precise predictions. The method combines the capacity to fast store and retrieve data from external memory with the gradual learning of an abstract gradient-based method for producing meaningful representations of raw data. A new meta-learning technique called memory-augmented meta-learning on the meta-path addresses the cold-start recommendation on the meta-path (Li et al., 2022). The approach uses memory-augmented meta-optimization to keep the meta-learning model out of the local optimum and builds at the data level to enhance the pertinent semantic information of the data. In order to obtain generalizable knowledge that can be applied to new tasks, meta-learning techniques often entail training a model on a range of diverse tasks. Memory-augmented meta-learning boosts generalisation and speed, resulting in learning architectures that outperform manually created models.

A type of memory-augmented model called a neural Turing machine (Graves et al., 2014) was developed as an inspiration for Turing machines. They integrate a controller for a neural network with a read-write external memory bank. NTMs have an addressing mechanism that facilitates effective content- and location-based information retrieval and storage. They have been applied to meta-learning to enhance the model's capacity for memory and improve its ability to generalize to new tasks.

Architecture: Memory-augmented meta-learning architectures combine the concept of external memory with the meta-learning framework. These architectures typically consist of a controller (such as a recurrent neural network) that interacts with the external memory module. During meta-training, the memory is used to store task-specific information or gradients, enabling faster adaptation to new tasks during meta-testing. The memory-augmented models can read from the memory to access relevant information and write to the memory to update or store new information. These architectures provide enhanced memory capacity and retrieval mechanisms to improve meta-learning performance (Santoro et al., 2016).

Memory-augmented meta-learning has shown promising results in tasks that require memory retention, reasoning, or complex decision-making. By incorporating external memory components, these models can effectively store and retrieve information across tasks, facilitating faster adaptation and better generalization. However, designing efficient reading and writing mechanisms, managing memory allocation, and addressing scalability and computational challenges remain areas of ongoing research.

1.6 Evolutionary Strategies

The principles of natural evolution serve as the foundation for a family of optimisation algorithms known as evolutionary algorithms (EAs). They are frequently employed to resolve challenging optimisation issues when conventional gradient-based approaches may falter. In order to find the best answer, EAs often keep a population of candidate solutions and employ genetic operators like selection, crossover, and mutation to evolve the population across generations. Evolutionary Strategies (ES) is a specific class of evolutionary algorithms that focus on optimizing continuous-valued problems. ES algorithms employ a population of candidate solutions, and instead of using traditional genetic operators, they use mutation-based search and update the population based on the performance of individuals. ES methods often leverage techniques like covariance matrix adaptation and evolution paths to guide the search towards promising regions in the solution space.

Evolutionary Meta-Learning combines the principles of meta-learning and evolutionary algorithms. It seeks to enhance the meta-learning procedure itself, for example, by identifying appropriate initializations or modifying meta-parameters.

The meta-parameters or designs of meta-learners are evolved using evolutionary algorithms, enhancing their generalisation and task-specific adaptability.

1.7 Applications of Optimization Approaches in Meta-Learning

Optimization-based meta-learning approaches have several applications in machine learning. Following are a few examples of optimization-based meta-learning applications:

1) Quick learning of new tasks: Meta-learning's objective is to assist the model in swiftly adapting to or picking up new skills based on a small number of examples. Model Agnostic Meta-Learning (MAML), for example, finds ideal parameters that can be adjusted to perform well on new tasks using optimization-based meta-learning techniques.

2) Few-shot learning: For few-shot learning, which entails learning from a constrained set of examples, optimization-based meta-learning algorithms can be utilised. Meta-learning techniques focused on optimisation can help the model learn more quickly and generalise to new tasks more effectively by learning a good initialization of the model parameters (Ravi & Larochelle, 2017).

3) Transfer Learning: Transfer learning involves leveraging knowledge learned from a source domain to improve learning in a target domain. Optimization techniques can be used in transfer learning to adapt the parameters of a pre-trained model to the target domain. By optimizing the model's parameters through fine-tuning or other transfer learning methods, the model can be adjusted to perform well in the target domain using a limited amount of labeled data.

4) Domain Adaptation: Domain adaptation focuses on adapting a model trained on a source domain to perform well on a different target domain. Optimization methods can be used to minimize the distribution shift between the source and target domains. This involves optimizing the model's parameters to align the feature distributions or reduce the domain discrepancy, allowing the model to generalize effectively to the target domain.

5) Reinforcement Learning: Reinforcement learning (RL) involves training agents to make sequential decisions in an environment to maximize a reward signal. Optimization techniques play a crucial role in RL, where algorithms like policy gradient methods and Q-learning use optimization methods to update the agent's policy or value function parameters. Meta-learning approaches can be employed to optimize RL algorithms, enabling faster learning, better exploration, and improved adaptation to new tasks or environments.

Optimization approaches can also be applied to optimize the learning algorithms themselves. Meta-learning can be used to learn hyperparameters, architectures, or optimization strategies that improve the learning process. By optimizing the learning algorithm's parameters, such as learning rates, regularization coefficients, or network architectures, meta-learning can enhance the efficiency and effectiveness of the learning process, leading to improved performance on a wide range of tasks.

REFERENCES

Bian, W., Chen, Y., Ye, X., & Zhang, Q. (2021). An Optimization-Based Meta-Learning Model for MRI Reconstruction with Diverse Dataset. *Journal of Imaging*, *7*(11), 231. doi:10.3390/jimaging7110231 PMID:34821862

Chaehan, S. (2021), Exploring Meta Learning: Parameterizing the Learning-to-learn Process for Image Classification. *International Conference on Artificial Intelligence in Information and Communication (ICAIIC)*. IEEE.

Correa Silva, A. (2020). Meta-Learning Applications in Digital Image Processing. *Proceedings of International Conference on Systems, Signals and Image Processing (IWSSIP)*, (pp 19-20). IEEE.

Finn, C., Xu, K., & Levine, S. (2019) *Probabilistic Model-Agnostic Meta-Learning. 32nd Conference on Neural Information Processing Systems (NeurIPS 2018)*, Montréal, Canada

Grant, E., Finn, C., Peterson, J., Abbott, J., Levine, S., Darrell, T., & Griffiths, T. (2017) Concept acquisition through meta-learning. In *NIPS Workshop on Cognitively Informed Artificial Intelligence*. IEEE.

Finn, C., Abbeel, P., & Levine, S. (2017). Model-agnostic meta-learning for fast adaptation of deep networks. In *International Conference on Machine Learning (ICML)*. IEEE.

Graves, A., Wayne, G., & Danihelka, I. (2014). Neural Turing Machines, Neural and Evolutionary Computing. arXiv. https://doi.org/ doi:10.48550/arXiv.1410.5401

Hospedales, T., Antoniou, A., Micaelli, P., & Storkey, A. (2022). Meta-Learning in Neural Networks: A Survey. *IEEE Transactions on Pattern Analysis and Machine Intelligence*, *44*(9). PMID:33974543

Huisman, M., van Rijn, J. N., & Plaat, A. (2021). A survey of deep meta-learning. *Artificial Intelligence Review*, *54*(6), 4483–4541. doi:10.100710462-021-10004-4

Khan, I., Zhang, X., Rehman, M., & Ali, R. (2020). *A Literature survey and empirical study of Meta-Learning for Classifier Selection* (Vol. B). IEEE.

Khodak, M., Balcan, M. F., & Talwalkar, A. (2019) Provable Guarantees for Gradient-Based Meta-Learning. *Proceedings of the 36 th International Conference on Machine Learning, Long Beach, California*, 97. IEEE.

Li T., Su, Xin., Liu, Su., Liang, W., Hsieh, M.Y., Chen, Z., Liu, X.C., & Zhang, H. (2022) *Memory-augmented meta-learning on meta-path for fast adaptation cold-start recommendation.* Taylor and Francis.

Li, Z., Zhou, F., Chen, F., & Li, H. (2017) Meta-SGD: Learning to learn quickly for few-shot learning. arXiv

Park, H., Lee, G., Kim, S., Ryu, G., Jeong, A., Park, S., & Sagong, M. (2021) A Meta-Learning Approach for Medical Image Registration. *IEEE 19th International Symposium on Biomedical Imaging (ISBI)*. IEEE.

Ravi, S., & Larochelle, H. (2017) Optimization as a Model for Few-Shot Learning. *International Conference on Learning* . IEEE

Santoro, A., Bartunov, S., Botvinick, M., Wierstra, D., & Lillicrap, T. (2016). *Meta-Learning with Memory-Augmented Neural Networks.* Proceedings of the 33rd International Conference on Machine Learning, New York, NY.

Sun, J., & Li, Y. (2021). MetaSeg: A survey of meta learning for image segmentation. *Cognitive Robotics.*, *1*, 83–91. doi:10.1016/j.cogr.2021.06.003

Yao, H., Huang, L. K., Zhang, L., Wei, Y., Tian, L., Zou, J., & Huang, J. (2021) Improving generalization in meta-learning via task augmentation. In *Proceedings of the International Conference on Machine Learning, Virtual Event*. IEEE.

Ye, J., H., Chao, W., (2022) How to train your MAML to excel in Few-Shot Classification. *The International Conference on Learning Representations (ICLR)*. IEEE.

Chapter 2
A Review on Image Super-Resolution Using GAN

Ajay Sharma

🆔 https://orcid.org/0000-0001-7951-9371
VIT Bhopal University, India

Bhavana Shrivastava
Maulana Azad National Institute of Technology, India

Swati Gautam
Maulana Azad National Institute of Technology, India

ABSTRACT

This study focuses on the utilization of generative adversarial networks (GANs) for generating high-resolution facial images from low-resolution inputs, which is vital for computer vision applications. Facial images present a complex structure, posing challenges for obtaining high-quality results using traditional super-resolution methods. However, recent advancements in deep learning, particularly GANs, have shown promising outcomes in this area. In this work, the authors conduct a comprehensive analysis of state-of-the-art GAN-based techniques for realistic high-resolution face image generation. They discuss the principles of image degradation, the learning process of GANs, and the challenges associated with these methods. By offering insights into the current state and future research directions, they aim to familiarize readers with the context and significance of GAN-based face image generation. This work highlights the importance of GANs in improving facial image quality and their relevance to advancing computer vision applications such as face verification and recognition.

DOI: 10.4018/978-1-6684-7659-8.ch002

1. INTRODUCTION

Face hallucination super resolution is a domain specific task which generally used to resolve the problem of unimaginable facial image to imaginable facial images. This task is important because face recognition is important in surveillance purpose and in computer vision (Wang, Chen, Nie et al, 2020; Yu & Porikli, 2016; Zheng & Shao, 2018). If the image is present in small pixel values, then after magnification of image it becomes blurry so it is difficult to determine face image (Chen & Tong, 2017). Therefore, Baker et.al.(2000), proposed Gaussian pyramid model for predicting high frequency of HR image. To map the face image perfectly training is required and then deep learning introduced for training the network. Deep Learning is a process in which it works on CNN network for super resolving the image and un-blur the image by different CNN models but our mainly focused is on Generative Adversarial Network GAN (Goodfellow et al., 2017), UR-DGN (Yu & Porikli, 2016), ESRGAN (Yu, 2019), WGAN (Chen & Tong, 2017), TDAE (Xu,, 2017), LS GAN (Qi, 2017), Hi-face GAN (Yang & Liu, 2020) suggested. The magnification does by the different states of art methods up to 8x. The existing SR method is totally based on the quality of the given low-resolution image & the availability of high-resolution image. The process is only applicable if, facial features can be found in low resolution image and similar image is trained as a reference for the mapping. But, still there is a margin of error for the low-resolution tiny image, Typical variation in pose, illumination issue and ghosting artifacts in the reconstructed HR image. Existing state of art method with deep learning are failed to generate authentic result of facial details because many of the techniques are patch based which ignore the information of image class. When low resolution image is super resolved using 8x magnification using upscaling techniques 98.5% of the image information is lost (Jiang, 2016; Li et al., 2020; Zhi-Song, 2019). Despite the fact that the as of recent proposed CNN-based SR arrangements give state of arts quantitative outcomes as far as peak signal to noise ratio (PSNR) when they improve for recreation losses, for example, L1 or L2 in space, the outcomes are smooth without the fine subtleties required for good perceptual quality. This issue is more obvious with the expansion of the upscaling factor. On head of that, the PSNR measure can't catch perceptually significant contrasts between two pictures as it depends on the distinction between pixel level qualities at a similar position. One approach to bring perceptually significant element into SR picture is to utilize generative adversarial network. These organizations help to make reconstruct SR pictures that appear as though HR pictures, which are normally keener and contain fine information of image. This Facial image-based task received attention recently, there are few most popular studies of FHSR based on generative adversarial network is discussed in this chapter.

2. RELATED WORK

2.1 Generative Adversarial Network

According to author Good fellow et al. (2017), this model has a dual pair framework design for producing the image super resolution, making it a type of unsupervised learning model. This dual pair is also known as the discriminator network and the generative network. The generator's job is to gather noise and produce an approximate image that is closer to the truth, after which a discriminator is used to distinguish between the Real Image (RI) and the Generated Image (GI). To produce better image reconstruction, the discriminator and generator are each trained separately. If an HR image discriminator is used to predict an image with input k and an output F(k), and the probability that the discriminator's output is F(k) is 1, then the image is real image. Similarly, if the output generated from the discriminator is 0 then it is referred as Generated image which is nearer to the ground truth.

F(k) = 0, k = GI

1, k = RI

The disadvantage of this Generative Adversarial Network is that it will oscillate and experience gradient loss if the discriminator is not trained correctly.

When F(k) = 0.5, the discriminator has trouble determining whether an image is artificially generated or real. min max Ek ~ Ir x log F(k) + E k ~ Ig x log (1 − F(k)) (1)

G D

Loss functions of the discriminator is given by -

Loss(d) = − Ek~ Ir x logF (k) − E k ~ Ig x log (1 − F(k))   (2)

& Loss function of the generator is given by -

And Loss(g) = − Ex ~ Ig x log (1 − F(k))   (3)

where Ir and Ig are known as real data and the generated data.

To resolve the problem arises in the GAN model, many improved networks of Gan like WGAN (Chen & Tong, 2017) and LSGAN (Qi, 2017) were proposed. WGAN introduces Earthmover to calculate the data distribution between real and generated image respectively. It also resolves the issue of generative adversarial network which uses the KL divergence (Sharma, 2016) and Jensen Shannon divergence (Yu et al.,

Figure 1.
Source: Goodfellow et al. (2017)

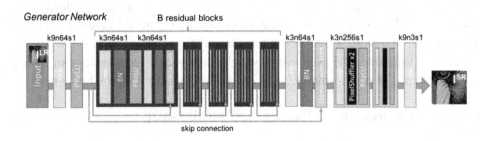

Figure 2.

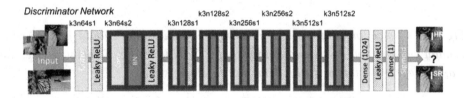

2018). Gan model becomes unstable when the real and generated data does not overlap completely. Then this approach WGAN shows the real distance between real and generated data and discriminator network tries to minimise the loss function.

$$\text{Loss(d)} = \text{Ek} \sim \text{Ig x F k} - \text{Ek}\sim \text{Ir x F k} \qquad (4)$$

and generator of WGAN tries to reduces the loss function as follow,

$$\text{Loss(g)} = - \text{Ek} \sim \text{Ig x F k} \qquad (5)$$

In LSGAN (Qi, 2017), they went against the sigmoid cross entropy loss used in generative adversarial network might allow the generator to convey substandard quality pictures whether or not the generator has confused the discriminator. In LSGAN (Qi, 2017) holds least square loss to drive the made tests to push toward the certifiable data tests. During the GAN' planning, the direct differentiations achieved two unmistakable misfortune capacities are showed up in Figure 1a and Figure 1b shows, while using the sigmoid cross entropy to choose the decision furthest reaches of the certified data and the fake data, whether or not some fake models for invigorating the generator are far from the contrasting authentic models, the decision botch is especially little as they are on the right half of as far as possible where the real data stays in also (the generator deceived the discriminator). LSGAN approach replaces the sigmoid cross-entropy misfortune with the least squares misfortune so it can stretch the fake guides to the decision edge as Figure 2a, Figure 2b, Figure 2c as shows. In a word, LSGAN can even more effectively drive the generator to create tests towards as far as possible, which can work on the adequacy of model improvement. The discriminator D of LSGAN attempts to limit the loss work.

Loss(d)=12Ek~Irx[(F(k)−b)2]+12Ek~Igx[(F(G(k))−a)2], (6)

Where generator loss is tries to minimise as follow,

Loss(g) =12Ek~IgxF G(k)−c2 (7)

Figure 3. Choice limits of two loss function. The choice limit should go across the genuine information dissemination for active GAN learning. Something else, the learning cycle is immersed which implies the slope is 0.

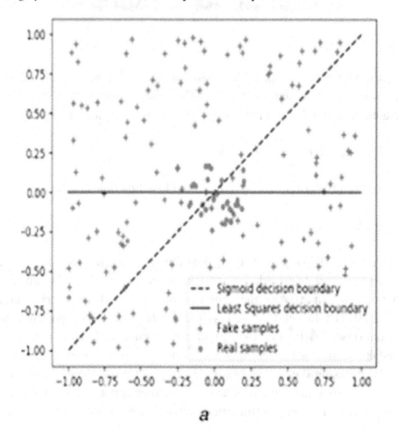

a

Figure 4. Choice limits of the sigmoid cross entropy misfortune work. It becomes little mistakes for the phony examples for stimulating G as they are on the right half of the choice limit.

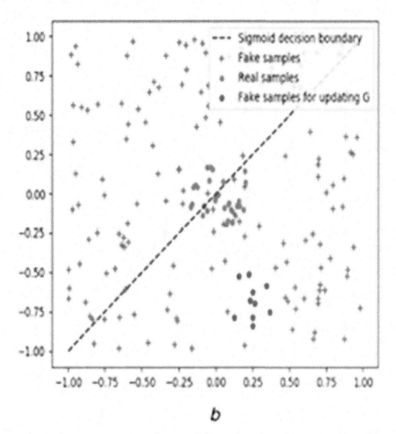

b

This outline is referred from (Chen et al., 2018). where the approximations of a, b and c are dependent upon the conditions b − a = 2 and b − c = 1. For example, a = − 1, b = 1 and c = 0.

2.2 Face Hallucination GAN Based

2.2.1 Face Hallucination Using UR-DGN

The network architecture of the proposed UR-DGN, proposed by Prokili et al. (2016), is used to resolve the small image of size (16*16), which is an improvement over the previous architecture of the original GAN introduced by Yu & Prokili, which was unable to provide satisfactory results on small images. There are also two networks

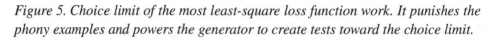

Figure 5. Choice limit of the most least-square loss function work. It punishes the phony examples and powers the generator to create tests toward the choice limit.

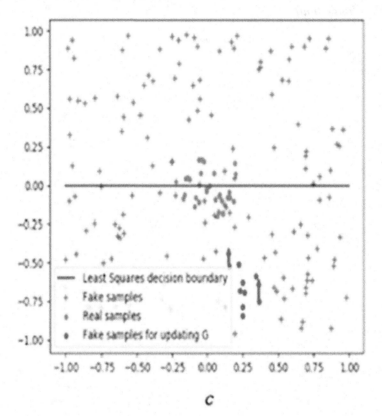

c

in the UR-DGN. both a discriminative network and a generative network. Due to the fact that it contains the only combination of deconvolution layers, the generative network differs from the original GAN. A discriminator is used to distinguish between the real image and the generated image during training, and the generator uses a Mean Square Error Loss Function for the supervised learning of image data sets. The URDGN pipeline process is depicted in Figure 3, where the full connected layer of deconvolution in the generator is displayed. This layer is used to map an image from beginning to end and to produce local details of face images, both of which aid in the reconstruction of an image with a higher PSNR. Equation illustrates how the MSE and adversarial loss function are added. The adversarial loss for this network's face image input ILR is The output probabilities of the discriminator F(G(ILR)) are a well-defined measure of the loss adversarial of generator G.

$$\text{Loss adversarial} = -\text{EILR} \sim \text{IL l log F G (ILR)}, \tag{8}$$

where Distribution of the low-resolution (LR) face images is represented by ILR, Mapping probability is represented by F (×) and G(ILR) is the generated high-resolution (HR) face image by the generator G. Then the loss function of G can be denoted as

$$lg = \lambda \, lMSE + ladv, \tag{9}$$

where λ is a weight to balance adversarial loss and MSE loss.
And the loss of discriminator D can be described as,

$$ld = - EIHR \sim IHR \, h \, logD \, IHR - EILR \sim IL \, llog \, 1 - F \, G \, (ILR), \tag{10}$$

where Distribution of the real high-resolution face images is represented by IHR represents real high-resolution face image.

2.2.2 Face Hallucination Using ESRGAN

A network was introduced by authors Wang et al. (2019). It addresses several problems with fine texture details in SRGAN. The ESRGAN network improves the discriminator network by using Realistic gan, which ensures that the generated image is more real than the other image, and it also introduces the VGG network to improve loss function before activation. Training is made easier with the increase in the capacity of training images thanks to the use of the ESRGAN network. To distinguish between the real image (Xr) and the fake image, this architecture improved the generator and the discriminator section (Xf).In Gan network it uses a standard discriminator network which tries to predict the probabilistic image between X_r and X_f is shown in Figure 4 The standard discriminator is expressed as,

$F(x) = \sigma(C(x))$, where σ is referred as sigmoid function and $C(x)$ is referred as non-transformed discriminator.

Then the Relativistic Discriminator is expressed as,

$F (X \, r, X_f) = \sigma (C (Xr) - Ex_f [C (_x f)])$, where Exf (.) is the mean data of mini batch and the loss function of the discriminator function is formulated as:

$$LD = -E_{xf} [log (F (Xr, Xf)_)] - Exf [log (1 - F (X \, r, Xf)_) \tag{1_1}$$

Adversarial generator loss is also formulated as,

$$LG = \_Exr [l_og (1 - F (Xr, X_f))] - Exr [l_og (F (X \, r, X_f) \tag{_12}$$

This modification gives more sharp edges and fine texture details.

Figure 6. Ur-Dgn: The generator network (with red block)
Source: This diagram is reproduced from Yu and Porikli (2016)

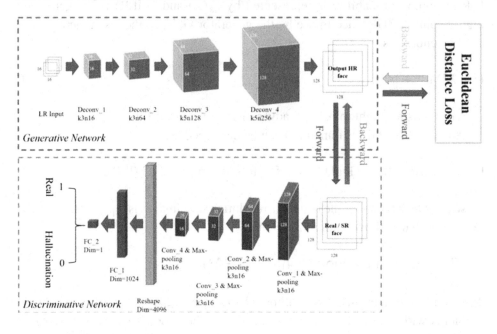

Figure 7. ESRGAN
Source: This diagram is reproduced from Yu (2019)

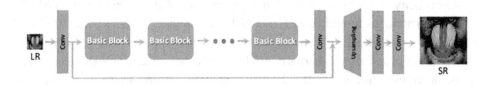

2.2.3 Face Hallucination Super Resolution Using Hi Face GAN

The network architecture of Hi face Gan (Yang & Liu, 2020) introduces a Collaborative suppression and replenishment (CSR) network to overcome the restoration of fine texture details in GAN(Goodfellow et al., 2017), UR-DGN(Yu & Porikli, 2016) and ESRGAN(Yu, 2019). The CSR network consist of two cascaded networks of suppression module and replenishment module is shown in Fig. 5. where, suppression module is used for generating high resolution image reconstruction from heterogeneous degradation where, CNN Layers suffer from discriminating image contents. Comparison with loss function, hi face Gan has convincing visual details such as hair bangs, beard, wrinkles and photo realistic results which shows

Table 1. Represents the algorithm of UR-DGN and ESRGAN

Algorithm 1 Mini Batch Stochastic Gradient Decent Training of UR-DGN	Algorithm 2 Mini Batch Stochastic Gradient Decent Training of ESRGAN
Input: Mini batch size N, LR and HR face image pairs (ILR$_i$,IHR$_i$),	Input: Mini batch size N, LR and HR face image pairs (X$_r$, X$_f$),
maximum number of iterations K	maximum number of iterations K
1. while iteration < K	1. while iteration < K
2. Choose one mini batch of LR and HR image pairs (ILR$_i$, IHR$_i$), i= 1,....,N	2. Choose one mini batch of LR and HR image pairs (X$_r$, X$_f$) r = 1,..,N, f = 1,....,N
3. Generate one minibatch of HR face images from G (ILR)	3. Generate one minibatch of HR face images from G (X$_r$)
4. Update the parameter of Discriminator	4. Update the parameter of Discriminator
5. Update the parameter of Generator	5. Update the parameter of Generator
6.end while	6.end while
Output: URDGN	**Output: ESRGAN**

the better performance of hi face gan over ESRGAN and Super Fan. Most of the face images are blurry due to insufficient amount of detail so to avoid this problem it follows the adversarial loss scheme and the loss function used by this approach is formulated as:

$$L^{GAN} = E[\left\|\log F\left(Ig-1\right)\right\|_f^2] + E[\left\|\log F\left(Ig\right)\right\|_f^2] \qquad (13)$$

Where Ig is generated image.

The overall reconstructed image is the summation of the perceptual loss L_p, L^{GAN}, Matching Loss L_M.

$$L_{Reconstructed} = L^{GAN} + \lambda_p L_p + \lambda_M L_M \qquad (14)$$

Where λ_M and λ_p are the scaling factor for matching loss and perceptual loss respectively.

2.2.4 Face Hallucination Using TDAE Network

In previous studies all the state of art model for face images is based on noise free facial images and align images. But due to the interference of other disturbance present in environment introduces noise and un-alignment of image which do not generate high resolution image in a perfect manner so Author Yu et al(2017).,

Figure 8. Hi-Face GAN
Source: This diagram is reproduced from Yang and Liu (2020)

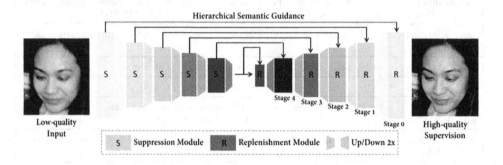

suggest a network which have three networks for upscaling the image for super resolving. A Combination of Transformative encoder and discriminative decoder is used (TDD). The TDD consist of two networks transformative upscaling and transformative discriminator The TDD network is used to upscale the image with the help of deconvolution layers and also it aligns the feature map of the face image by using spatial transformative network (STN). Due to the fact that low resolution image consists of noise so it is initially sent to the encoder to get aligned face image with high resolution face image the architecture is shown in Fig. 6, second decoder network is used to get the information of the hallucinated face image and it also uses MSE loss function for optimizing better result with compare to real image.

The loss function may be described as,

$$\text{Loss}_{TUN} = EI_{LR} \sim P_{Lr} 1, I_{HR} \sim P_{Hr} h \| G\, I_{LR} - I_{HR} \|^2_F \tag{15}$$

where $\| \times \|F$ means the Frobenius normalisation and the loss function of the discriminator which required to be minimised is the same as equation (9). The outputs of the decoder I^{SR} and the ground-truth aligned low resolution images I_{LR} are used to train the transformative encoder. The loss function of the encoder is denoted as,

$$\text{LEnc} = E\, I_{LR} \sim P_L\, L, I_{HR} \sim P_H\, h \| \Psi\, G\, I_{LR} - I_{HR} \|^2_F \tag{16}$$

$$\text{Loss (d)} = -\, EI_{HR} \sim P_{Hr}\, h\, [\log F\, (I_{HR}, a)] - EI_{LR} \sim P_{Lr}\, 1\, [\log 1 -$$

$$FG(I_{LR}, a) + \log 1 - D\, (I_{HR}, a))], \tag{17}$$

$$\text{Loss(g)} = EI_{LR} \sim P_{Lr}\, 1, I_{HR} \sim P_H\, h \| G\, I_{LR} - I_{HR} \|F^2 + \alpha\phi\, G$$

$$I_{LR} - \phi(I_{HR})F^2 - \beta\log F \ G(I_{LR}, a) \tag{18}$$

where $\Psi \times$is represented as the mapping function from the face hallucinated image $G \ I_{LR}$ from the LR face image I_{LR}.

and Distribution of the real high-resolution and low-resolution face images is represented by P_{Hr} and P_{Lr} respectively.

2.2.5 Face Hallucination Using Super Resolving Face Image With Supplementary Attributes

For Up sampling an image Auto encoder and deconvolution layer is used with multiple skip connections and Discriminator is used for super resolved images contains desired information of attributes of face images (Yu et al., 2018) is shown in Figure 7. This network is used to super resolve the images of 16x16 very tiny image to 8x magnification using upscaling and achieves superior results with compared to different state of art algorithms. The LR input is given to the auto encoder which consist of encoder, residual block and decoder with skip connections then the output is feed in the deconvolution for upscaling an image with filter size of 3x3 and then STN is used for alignment of the facial images and upscales using deconvolution layer after generating the up-sample image it is given to the input and discriminator used to differentiate between real or fake images.

Figure 9. TDAE
Source: This diagram is referred from Xu, (2017)

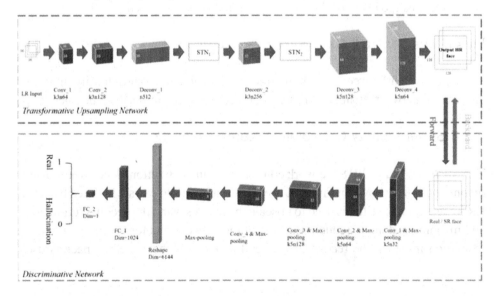

Table 2. Represents the algorithm of UR-DGN and ESRGAN

Algorithm 3 Mini Batch SGD Training of Hi Facegan	Algorithm 4 Mini Batch SGD Training of TDAE
Input: Mini batch size N, LR and HR face image pairs (I_r, I_g),	Input: Mini batch size N, LR and HR face image pairs (I_{LR}, I_{HR}),
maximum number of iterations K	maximum number of iterations K
1. while iteration < K	1. while iteration < K
2. Suppression Module after LR images (I_r). r = 1..,..,N	2. Suppression Module after LR images (I_{LR}). r = 1,.....,N
3. Generate minibatch of HR face images from Replenishment Module	3. Generate minibatch of HR face images from Replenishment Module
4. Upscale / Downscale 2x	4. Upscale / Downscale 2x
5.end while	5.end while
Output: Hi faceGAN	**Output: TDAE**

Discriminator loss function is given as –

L_d = - E [log D (h,a)] - E[log(1-D(h$_1$,a) + log(1-D(h,a$_i$)]

=-E[log D (h$_i$,a$_i$)]-E[log(1-D(h$_i$,a)-E[log(1-D(h,a$_i$)   (19)

Where, h$_i$ is HR super resolved face image, h is given as LR image, a is ground truth attributes, a$_i$ is mismatch attributes of face image, D (h$_i$,a$_i$)], D(h$_i$,a) and D(h,a$_i$) are the output of D.The loss function can be minimized by updating the networks and to reduces the Loss for the up-sampling network it is given as –

$Lu = E [\|h_i - h\|^2_{F+} \alpha \| \phi(h_{i)} - \phi(h) \|_F^2 - \beta log D (h_i, a)$ (20)

Where, weight between the feature similarity is denoted by α and intensity similarity and weight between attribute similarity and appearance similarity is denoted by β.

2.2.6 Super Identity Convolution Neural Network

The SICNN (Zhang, 2018) network consist of feature extraction, deconvolution and reconstruction is shown in Fig. 8. It uses dense network to extract features from the LR images and 3x3 filter is used to upscale the images using the deconvolution layer and mapping is done to reduces the cost of the network and for the prediction high resolution image at the reconstruction end and hypersphere metric space is used for the better identity representation. The Loss function used by SICNN is super resolution loss and is given by,

Super Resolution Euclidean loss is given as.

$$SuperResolution_{loss} = CNN_H \left(I_j^{LR} \right) - I_j^{HR2}{}_2 \tag{21}$$

Where, I_j^{LR}, I_j^{HR} = i-th low-resolution and high-resolution image respectively and $CNN_H \left(I_j^{LR} \right)$ = output of network with input I_j^{LR}.

This algorithm is used to overcome the dynamic domain divergence problem.

The training procedure is new which trained alternately in each iteration because baseline technique generates a dynamic divergence problem.

3. COMPARISONS AND DISCUSSIONS

A comprehensive assessment for the Gan-based networks for face representation strategies appears in Table 4. From the Table, Celeb A is the most applied preparing dataset and Torch is the most well-known advancement step for these methodologies. Then, considering the way that low resolution face images are reliably un-aligned, unrestrained TDAE (Xu,, 2017), FaceAttr (Yu et al., 2018) will show the more grounded pertinence. The presentation associations for all these Gan-set up face

Table 3. Represents the algorithm of UR-DGN and ESRGAN

Algorithm 5 Mini Batch SGD Training of Hi faceGAN	Algorithm 6 Mini Batch SGD Training of SICNN
Input: Mini batch size N, LR and HR face image pairs (h, h$_i$),	Input: Face recognition model CNNR trained by HR facial images, face hallucination model CNNH trained by L SR, minibatch size N, LR and HR facial image pairs I LR i, IHR i .
maximum number of iterations K	
1. while iteration < K	1. while not converge do
2. Suppression Module after LR images (I$_r$). r = 1,..,N	2. By Choosing 1 minibatch of N LR and HR image pairs (I_j^{LR}, I_j^{HR}), j = 1, ..., N.
3. Generate minibatch of HR face images from Replenishment Module	3. Generate one minibatch of N hallucinated facial images I_j^{HR} from I_j^{LR} j = 1, ..., N, where I_j^{HR} = CNNH (I_j^{LR}).
4. Upscale / Downscale 2x	4. Update the recognition model by descending its gradient.
5.end while	6.end while
Output: Face Attr.	**Output:SICNN**

Figure 10. Face image with supplementary attributes
Source: This diagram is referred from Yu et al. (2018)

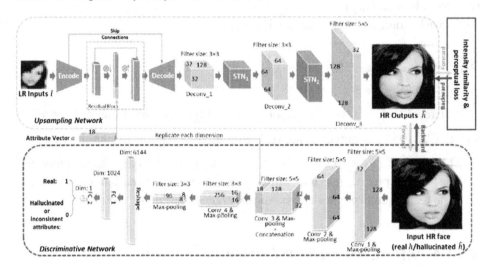

Figure 11. SICNN
Source: This diagram is referred from Zhang (2018)

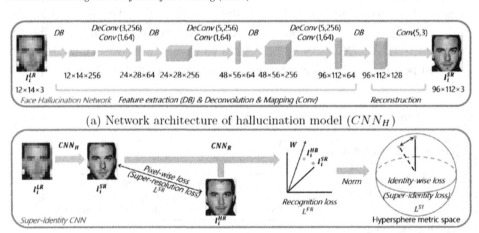

representation arrangements based about the mutual place psnr and ssim measures of the images are shown in Table 5. For sure, these Gan based strategies adjacent to Super-fan, 18 thousand face pictures of Celeb A (Zhang et al., 2019) dataset is utilized for preparing and essentially one more 2 thousand Celeb A photograph are genuinely utilized for execution testing. Concerning Super-fan, it randomly picked 200 low resolution and actively confused face pictures from dataset for passionate evaluation. As per Tables 1, it is not difficult to draw that SICNN (Zhang, 2018) will

Table 4. The comparison between face hallucination techniques

Model	Learning Approach	Generator of Loss Function	Dataset for Training	Modeling Platform	Upscaling Factor	Unaligned Model
URDGN(Yu & Porikli, 2016)	rms Prop.	Mse+Cross Enotropy Loss	Celeb A	Torch – 7	8	X
ESRGAN(Yu, 2019)	rms Prop.	Mse+Adversarialoss	Celeba, Helen, Lfw	Torch – 7	8	✓
Hi-Face Gan(Yang & Liu, 2020)	rms Prop.	Mse+Cross Enotropy Loss+Perceptual Loss	Celeb A	Torch – 7	8	✓
TDAE (Xu,, 2017)	rms Prop.	Mse+Cross Enotropy Loss	Celeb A	Torch – 7	8	✓
Face Attr.(Yu et al., 2018)	rms Prop.	Mse+Cross Enotropy Loss+Perceptual Loss	Celeb A	Torch – 7	8	✓
SICNN(Zhang, 2018)	rms Prop.	Super Resolution Loss	CelebA, Helen	Torch – 7	8	✓

give the best settled execution with 8 × down-scaling factor. Besides, when going to 4 × down-scaling, ESRGAN (Yu, 2019) will change into the victor. Notwithstanding giving various models to accomplish distinctive picture SR impacts, there are different factors unmistakably identified with the exceptional insight or the picture thought of the super-settled face pictures. One tremendous factor is the picture quality appraisal rule, as PSNR (top sign to-rattle degree) and root mean square error. In all honesty, these examination measures handle the general picture, yet it can't mirror the possibility of the subtleties that need thought in the stable picture. The SSIM list revolves around the secret comparability of two pictures. Notwithstanding, because of the amazing principal attributes of human face pictures, SSIM isn't reasonable for quality measurement of imagined face pictures. The indecency of these examination measures can in like way be seen from paying little brain to how high the assessment respect is, essentially facial data is going be lost all considered the settled face pictures made by all these Gan-based super resolution strategies. Consequently, tracking down an appropriate and all-around focused quality evaluation for manufactured facial pictures will be a hopeful and basic examination heading for face pipedream. Additional fundamental issue is dataset utilized for testing and preparing. Initial of face area is non-general fittingness of the current face super resolution methods when opposed with various datasets. In spite of the fact that a massive part of the proposed face super resolution techniques reports staggering execution, the helpful settled impact of these strategies relied genuinely on important datasets. Face image super resolution techniques (Fan et al., 2020; Hayat, 2018; Ji, 2020) that perform fine in different and complex categories of magnification and alliances, as a general rule, low resolution conditions are yet to be made. In like way, most current frameworks utilize planned LR information that debased from the differentiating HR information ()[24]. with predefined degradation models (like Bicubic). In any case, concerning the novel and difficult nature of destructions, the predefined model can't mirror the reasonable applications. This raises the requirement for the standard and all-around datasets for face excursion. In this manner, making an instructive record that satisfies the dependable sight closed-circuit camera feeds will be of imperativeness. Also, most of the current works use face super resolution as a preprocessing step going before the face authentication or face technique (Zhang et al., 2019)[26]. Unquestionably, joint learning two endeavors could shockingly work on the introduction of both SR and face changed. Thusly, organizing a perform different undertakings learning association may be a promising and basic assessment heading for face depiction. Finally, by temperance of the incredible computational unpredictability, masterminding GANs regularly requires a long time and a similar figuring chip, for instance, GPU is required. Also, to clean everything off, it can hardly pass on the learned immense model doubtlessly on the introduced contraption due to the enormous number of model cutoff focuses with the drowsy getting speed. Thusly, building a basic learning unit a few cutoff focuses and fast affirmation speed, as electrifying learning, will be a future assessment bearing Some evaluation result of the techniques is shown in Fig. 9.

4. CONCLUSION

In this study, we looked at the newest, most well-known GAN-based face imagination strategies. The first GAN-based strategies, the more advanced GAN-based strategies, and GAN with expansion data-based ones can all be grouped together according to the specific models, as shown above. We discovered through a thorough analysis and discussion that these Gan-based strategies perform better when it comes to face remaking than traditional picture SR approaches. However, taking everything into

Table 5. The comparison between PSNR and SSIM

Model	Gan (Goodfellow et al., 2017)	ESRGAN (Yu, 2019)	URDGN (Yu & Porikli, 2016)	TDAE (Xu, 2017)	SICNN (Zhang, 2018)	Hi-Face Gan (Yang & Liu, 2020)	Face Attr. (Yu et al., 2018)
Testing dataset	CelebA	CelebA, Helen	CelebA	CelebA	CelebA, Helen	CelebA	CelebA
PSNR	20.384	21.001	24.725	20.42	26.894	23.705	20.536
SSIM	0.543	0.576	0.6926	0.56	0.768	0.619	0.540

account, most facial definite provisions will be missing in their captured super-settled face images. Finally, we look at a few related factors relating to face pipedream and offer some encouraging examination bearings. We acknowledge that this overview will help various specialists comprehend the characteristics and the development on face based imagination.

REFERENCES

Baker, S., & Kanade, T. (2000). Hallucinating faces. *Proc. Fourth IEEE Int. Conf. on Automatic Face and Gesture Recognition.* 10.1109/AFGR.2000.840616

Chen, Y., Tai, Y., Liu, X., Shen, C., & Yang, J. (2018). FSRNet: End-to-End Learning Face Super-Resolution with Facial 23. Priors. In *Proceedings of the IEEE Computer Society Conference on Computer Vision and Pattern Recognition* (pp. 2492–2501). IEEE. 10.1109/CVPR.2018.00264

Chen, Z., & Tong, Y. (2017). *Face super-resolution through Wasserstein Gans.* arXiv preprint arXiv.02438.

Fan, Z., Hu, X., Chen, C., Wang, X., & Peng, S. (2020, December). Facial image super-resolution guided by adaptive geometric features. *EURASIP Journal on Wireless Communications and Networking, 2020*(1), 149. doi:10.118613638-020-01760-y

Goodfellow, I., Pouget-Abadie, J., & Mirza, M. (2017). Generative adversarial nets. *Proc. Conf. and Workshop on Neural Information Processing Systems.*

Hayat, K. (2018). Multimedia super-resolution via deep learning: A survey. *Digital Signal Processing: A Review Journal, 81.* . doi:10.1016/j.dsp.2018.07.005

Ji, S. (2020). Kullback-Leibler Divergence Metric Learning. *IEEE Transactions on Cybernetics.* PMID:32721911

Jiang, J. (2016). Noise Robust Face Image Super-Resolution Through Smooth Sparse Representation. *IEEE Transactions on Cybernetics.* PMID:28113611

Li, J., Zhou, Y., Ding, J., Chen, C., & Yang, X. (2020). ID Preserving Face Super-Resolution Generative Adversarial Networks. *IEEE Access : Practical Innovations, Open Solutions, 8*, 1. doi:10.1109/ACCESS.2020.3011699

Mun, J., & Kim, J. (2020, November). Universal super-resolution for face and non-face regions via a facial feature network. *Signal, Image and Video Processing, 14*(8), 1601–1608. doi:10.100711760-020-01706-3

Qi, G.-J. 'Loss-sensitive generative adversarial networks on Lipschitz densities', arXiv preprint arXiv.06264, 2017.

Sharma, P. K. (2016). Dilation of Chisini-Jensen-Shannon Divergence. In *IEEE International Conference on Data Science and Advanced Analytics (DSAA)*. IEEE.

Xu, X., & Porikli, F. (2017). Hallucinating very low-resolution unaligned and noisy face images by transformative discriminative autoencoders. *Proc. Int. Conf. on Computer Vision and Pattern Recognition*.

Wang, L., Chen, W., Yang, W., Bi, F., & Yu, F. R. (2020). A State-of-the-Art Review on Image Synthesis with Generative Adversarial Networks. In IEEE Access (Vol. 8). Institute of Electrical and Electronics Engineers Inc. doi:10.1109/ACCESS.2020.2982224

Wang, Q., Chen, M., Nie, F., & Li, X. (2020, January 1). Detecting coherent groups in crowd scenes by multiview clustering. *IEEE Transactions on Pattern Analysis and Machine Intelligence*, *42*(1), 46–58. Advance online publication. doi:10.1109/TPAMI.2018.2875002 PMID:30307858

Yang, L., & Liu, C. (2020). *HiFaceGAN: Face Renovation via Collaborative Suppression and Replenishment*. doi:10.1145/3394171.3413965

Yang, W., Zhang, X., Tian, Y., Wang, W., Xue, J., & Liao, Q. (2019). Deep Learning for Single Image Super-Resolution: A Brief Review. *IEEE Transactions on Multimedia*. doi:10.1109/TMM.2019.2919431

Yu, K. (2019). ESRGAN: Enhanced Super-Resolution Generative Adversarial Networks. In *European Conference on Computer Vision*. Springer, .

Yu, X., Fernando, B., & Hartley, R. (2018). Super-resolving very low-resolution face images with supplementary attributes. *Proc. of the IEEE Conf. on Computer Vision and Pattern Recognition*. 10.1109/CVPR.2018.00101

Yu, X., & Porikli, F. (2016). Ultra-resolving face images by discriminative generative networks. *European Conf. on Computer Vision*, Amsterdam, Netherlands. 10.1007/978-3-319-46454-1_20

Zhang. (2018). Super-identity convolutional neural network for face hallucination. *Proc. Eur. Conf. Comput. Vis. (ECCV)*. IEEE. 10.1007/978-3-030-01252-6_12

Zhang, H., Wang, P., Zhang, C., & Jiang, Z. (2019, July). A comparable study of CNN-based single image super-resolution for space-based imaging sensors. *Sensors (Basel)*, *19*(14), 3234. doi:10.339019143234 PMID:31340511

Zheng, F., & Shao, L. (2018). A winner-take-All strategy for improved object tracking. *IEEE Transactions on Image Processing*, *27*(9), 4302–4313. doi:10.1109/TIP.2018.2832462 PMID:29870349

Zhi-Song, L. (2019). *Reference Based Face Super-Resolution*. IEEE.

Chapter 3
Optimizing Hyper Meta Learning Models:
An Epic

G. Devika

iD https://orcid.org/0000-0002-2509-2867
Government Engineering College, Krishnarajapete, India

Asha gowda Karegowda

iD https://orcid.org/0000-0002-1353-4293
Siddaganga Institute of Technology, India

ABSTRACT

Optimizing hyper meta learning models is a critical task in the field of machine learning, as it can improve the performance, efficiency, and scalability of these models. In this chapter, the authors present an epic overview of the process of optimizing hyper meta learning models. They discuss the key steps involved in this process, including task selection, model architecture selection, hyperparameter optimization, model training, model evaluation, and deployment. They also explore the benefits of hyper meta learning models and their potential future applications in various fields. Finally, they highlight the challenges and limitations of hyper meta learning models and suggest future research directions to overcome these challenges and improve the effectiveness of these models.

INTRODUCTION

Optimizing meta-learning models involves tuning the hyper parameters of the model to improve its performance on a given task. Meta-learning is a subfield of machine

DOI: 10.4018/978-1-6684-7659-8.ch003

learning that focuses on developing algorithms that can learn from experience to solve new problems quickly and efficiently(Chelsea et al, 2022). Meta-learning models are typically trained on a set of related tasks, and then used to quickly adapt to new tasks by leveraging the knowledge gained from the training tasks.

To optimize meta-learning models, it is important to select appropriate hyper parameters for the model, such as the learning rate, regularization strength, and network architecture. There are several techniques that can be used for hyperparameter optimization, including grid search, random search, Bayesian optimization, and gradient-based optimization (Yikai et al, 2019).

Grid search involves defining a set of possible values for each hyperparameter and testing all possible combinations of these values to find the combination that yields the best performance on a validation set. Random search involves randomly sampling hyperparameters from their defined distributions and evaluating their performance on a validation set (Adam et al, 2022). Bayesian optimization is a more advanced technique that uses Bayesian inference to build a probabilistic model of the objective function (i.e., the model performance) and select hyperparameters that are likely to yield the best performance. Gradient-based optimization involves optimizing the hyperparameters using gradient descent, where the gradient of the objective function with respect to the hyperparameters is computed and used to update the hyperparameters. In addition to hyperparameter optimization, other techniques can be used to improve the performance of meta-learning models, such as data augmentation, model ensembling, and transfer learning. By optimizing meta-learning models, it is possible to develop algorithms that can quickly adapt to new tasks and improve their performance over time.

Solving a machine learning use case involves a number of steps, including understanding the problem, collecting and preparing the data, selecting an appropriate machine learning model, training and evaluating the model, and deploying the model in a production environment. Here is a high-level overview of the process:

a. Problem Definition: The first step in solving a machine learning use case is to define the problem you are trying to solve. This involves understanding the business objectives, the available data, and the potential impact of the solution.

b. Data Collection and Preparation: Once the problem is defined, the next step is to collect and prepare the data. This involves identifying the relevant data sources, cleaning and pre-processing the data, and selecting appropriate features.

c. Model Selection: After the data is prepared, the next step is to select an appropriate machine learning model. This depends on the type of problem and the characteristics of the data. Common types of machine learning models include supervised learning, unsupervised learning, and reinforcement learning.

d.   Training and Evaluation: Once the model is selected, the next step is to train the model on the data and evaluate its performance. This involves splitting the data into training and testing sets, training the model on the training set, and evaluating its performance on the testing set.

e.   Deployment: After the model is trained and evaluated, the final step is to deploy the model in a production environment. This involves integrating the model into the existing system, monitoring its performance, and making updates and improvements as needed.

Throughout the entire process, it is important to iterate and refine the solution based on feedback and performance metrics (Adam et al, 2022). By following these steps, it is possible to build a machine learning model that effectively solves the problem at hand.

OPTIMIZATION OF HYPER META LEARNING ALGORITHMS

Hyper meta learning algorithms are a class of machine learning algorithms that can quickly adapt to new tasks by leveraging prior experience from related tasks. Meta-learning, also known as "learning to learn," is a subfield of machine learning that focuses on developing algorithms that can learn from experience to solve new problems quickly and efficiently. Hyper meta learning algorithms take this a step further by optimizing the meta-learning process itself, which involves tuning the hyperparameters of the model to improve its performance on a given task. This can be done by training the model on a set of related tasks and then using techniques such as Bayesian optimization, gradient-based optimization, or evolutionary algorithms to find the optimal set of hyperparameters for the model (Adam et al, 2022).

Hyper meta learning algorithms have a wide range of applications, including computer vision, natural language processing, and reinforcement learning. They have been used to develop algorithms that can quickly adapt to new environments, learn from small amounts of data, and generalize to new tasks with high accuracy. Some examples of hyper meta learning algorithms include MAML (Model-Agnostic Meta-Learning), Reptile, and SNAIL (Sparse and Hierarchical Attention Networks for Learning to Learn). These algorithms have shown promising results in a variety of tasks, including image classification, object detection, and language modeling.

Overall, hyper meta learning algorithms have the potential to revolutionize the field of machine learning by enabling models to learn more quickly and efficiently, and by reducing the need for large amounts of training data (Yikai et al, 2019).

Below are the steps required to solve a machine learning use case and to build a hyper meta learning model is Figure 1 depicts a flow of steps involved in it and described in next section in detail in further sections.

Define the OBJECTIVE

Meta-learning is a subfield of machine learning that focuses on learning to learn. The objective of meta-learning models is to develop algorithms that can learn from experience and adapt to new tasks quickly and efficiently.

Meta-learning models aim to acquire knowledge and skills that enable them to learn more effectively and efficiently from new tasks or situations. This includes learning how to:

Figure 1. Flow of steps

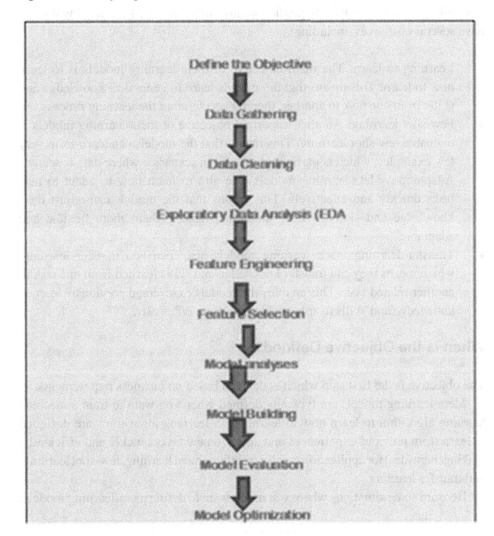

1. Generalize across tasks: Meta-learning models should be able to learn a high-level understanding of multiple tasks, so they can quickly adapt to new tasks that share similar characteristics.
2. Adapt to new tasks: Once a meta-learning model has learned a general understanding of a set of tasks, it should be able to quickly adapt to new tasks with minimal data (Chelsea et al, 2022).
3. Optimize learning: Meta-learning models should be able to optimize the learning process itself by selecting the best learning algorithms and hyperparameters for a given task

What's the Objective?

Deciding a use case you want to predict or know more about. Meta-learning models have several objectives, including:

*   Learning to learn: The main objective of meta-learning models is to learn how to learn. This means that the models learn to generalize knowledge and skills from one task to another, thereby accelerating the learning process.
*   Few-shot learning: Another important objective of meta-learning models is to enable few-shot learning. This means that the models can learn from very few examples, which is particularly useful in scenarios where data is scarce.
*   Adaptation: Meta-learning models also aim to learn how to adapt to new tasks quickly and effectively. This means that the models can adjust their knowledge and skills to new situations, making them more flexible and adaptable.
*   Transfer learning: Meta-learning models also focus on transfer learning, which means they can transfer knowledge and skills learned from one task to another related task. This enables the models to leverage previously learned knowledge and skills to solve new tasks more efficiently.

When is the Objective Defined?

The objective is the first step which is decided based on business requirements.
Meta-learning models are typically defined when you want to train a machine learning algorithm to learn how to learn. Meta-learning algorithms are designed to learn from previous experiences and adapt to new tasks quickly and efficiently, making them ideal for applications such as reinforcement learning, few-shot learning, and transfer learning.
Here are some situations where you might want to define meta-learning models:

a.   Few-shot learning: If you have limited labeled data for a new task, meta-learning algorithms can be used to learn from a few examples and generalize to new examples.

b.   Reinforcement learning: Meta-learning can be used to learn optimal policies quickly and efficiently by using past experiences to adapt to new tasks.

c.   Transfer learning: Meta-learning models can learn from past experiences and adapt to new tasks more efficiently than traditional transfer learning approaches.

d.   Domain adaptation: Meta-learning models can learn to adapt to new domains by leveraging past experiences.

Why is it Necessary to Set an Objective?

Defining the objective sheds light on what kind of data should be gathered. It also helps us in judging what kind of observations are important while doing exploratory data analysis.

Meta learning is a type of machine learning that focuses on learning how to learn. Meta learning models aim to develop strategies and techniques to optimize the learning process of other machine learning models.

Setting objectives for meta learning models is crucial for several reasons:

a.   Guiding the learning process: Setting objectives provides a clear direction for the meta learning model to follow. It enables the model to focus on specific learning tasks, which can help it to develop effective strategies for solving those tasks.

b.   Measuring success: Objectives provide a way to measure the success of the meta learning model. By defining clear objectives, it is possible to assess how well the model is performing and whether it is achieving the desired outcomes.

c.   Adapting to new situations: Objectives can be used to train meta learning models to adapt to new situations. By setting objectives that vary in difficulty, the model can learn to generalize its strategies and adapt them to new learning tasks.

How to Define an Objective?

An objective should be clear and precise. Therefore, to define a clear objective we need to follow few steps like:

• Understand the business (Eg: Grocery store)
• Identify the problem (Eg: Less Profits)

- List out all the possible solutions to solve the problem(Eg: By increasing sales or by reducing manufacturing costs or by managing inventory etc.)
- Decide on one solution (Eg: managing inventory, we can come to this conclusion by talking to the respective business people back and forth.

By following the above steps, we've clearly defined that the objective is to build a model to manage inventory in order to **increase** store **profits**.

Meta-learning models are typically designed to learn how to learn, or in other words, to learn how to effectively adapt to new tasks or problem domains quickly. Defining clear objectives for these models is essential to ensure that they learn the right things and can perform well on new tasks.

Here are some general guidelines for defining objectives for meta-learning models:

a.   Identify the domain or task space: First, you need to identify the domain or task space that you want the meta-learning model to operate in. This can be a broad domain (e.g., computer vision) or a specific task (e.g., object recognition).

b.   Determine the types of tasks: Next, you need to determine the types of tasks that the meta-learning model will encounter. These tasks should be representative of the domain or task space and should be diverse enough to capture the range of possible tasks that the model may encounter.

c.   Define the performance metric: You should define a performance metric that the meta- learning model will use to evaluate its performance on each task. This metric should reflect the objective that you want the model to achieve, whether it is accuracy, speed, or some other measure.

d.   Define the meta-objective: The meta-objective is the objective that the meta-learning model will optimize during training. It should reflect the overall goal of the meta-learning model, which is to learn how to quickly adapt to new tasks. Typically, this is done by minimizing the expected loss on new tasks or maximizing the expected performance on new tasks.

e.   Design the training procedure: Finally, you should design the training procedure to optimize the meta-objective. This typically involves training the model on a set of tasks and evaluating its performance on a held-out set of tasks. The training procedure should be designed to encourage the model to learn useful generalizations that can be applied to new tasks.

DATA GATHERING

What's Data Gathering?

Data Gathering is nothing but collecting the data required as per the defined objective.

Data gathering for optimization involves collecting and analyzing data to improve a particular process or system. The goal is to identify areas that need improvement, understand how they can be optimized, and make changes to achieve better results.

The process of data gathering for optimization typically involves the following steps:

1. Determine the data needed: Decide what data needs to be collected to gain insights into the problem.
2. Collect the data: Collect data from various sources, such as surveys, interviews, and online tools.
3. Analyze the data: Use statistical and analytical methods to analyze the data and identify patterns and trends.
4. Identify opportunities for optimization: Based on the analysis, identify areas that need improvement and potential solutions.
5. Implement changes: Make changes to the process or system to optimize its performance.
6. Monitor and evaluate: Continuously monitor the process or system to ensure that the changes have been effective and evaluate the results.

Data gathering for optimization can be applied to various fields, such as business, manufacturing, healthcare, and transportation. By collecting and analyzing data, organizations can make informed decisions, improve efficiency, reduce costs, and enhance customer satisfaction

When Do We Gather Data?

Once the objective is defined, we will collect data. In optimization models, data is typically gathered at the beginning of the modeling process to inform the problem formulation and parameter estimation. The data can be obtained through various means such as surveys, experiments, observations, or through historical records.

The data collected is then used to estimate the values of model parameters such as coefficients, constraints, and objective function coefficients. The parameter estimation process is critical to the accuracy of the model, and it is often done through statistical techniques such as regression analysis, maximum likelihood estimation, or Bayesian inference.

Once the parameters have been estimated, the optimization model is then solved to obtain the optimal solution. The solution can be used to guide decision-making and inform policy recommendations. It is important to note that the model may need to be updated periodically as new data becomes available or as the problem changes over time.

Why is Data Gathering Necessary?

Without past data, we cannot predict the future, hence Data Gathering is necessary. In general a dataset is created by gathering data from various resources based on the objective. One of the reasons for gathering data from multiple resources is to get more accurate results i.e.,"The more the data, the more accurate the results will be".

Data gathering is necessary for optimization models because optimization models rely on mathematical algorithms to identify the best solution to a problem. These algorithms require data as inputs to make informed decisions.

In order to create an effective optimization model, it is essential to have accurate and comprehensive data on the problem being addressed. This data can include information on constraints, such as budget or time limitations, as well as variables that impact the outcome, such as customer preferences or market trends.

Without reliable data, the optimization model may produce inaccurate or suboptimal results. In some cases, the model may not even be able to provide a solution at all. Therefore, data gathering is a critical step in the optimization modeling process. Additionally, as optimization models are often used for decision-making purposes, it is important to ensure that the data used is current and relevant to the problem at hand. This may require ongoing data collection and analysis to ensure that the model remains accurate and effective over time.

How is Data Gathering Done?

Data can be collected in one of the following ways mentioned below:

1. API's (like Google, Amazon, Twitter, New York Times etc.)
2. Databases (like AWS, GCP etc.)
3. Open source (Kaggle/UCI Machine Learning Repositories etc.)
4. Web Scraping (Not recommended, as often it is considered as illegal)

The order of Defining the objective and Data gathering steps can be changed. Sometimes we will have the data handy and we need to define the objective later and sometimes we will decide the objective first and then we will gather data.

Meta-optimization learning models involve training a meta-learner to find the best optimization algorithm for a given task. In order to do this, the model requires data to train on, which is typically collected through a process called "meta-dataset generation".

Meta-dataset generation involves creating a dataset of optimization problems, where each problem is paired with the best optimization algorithm to solve it. This dataset is then used to train the meta-learner to predict the best algorithm for a new problem.

The process of generating the meta-dataset can vary depending on the specific application and domain. In some cases, the problems and corresponding algorithms may be generated synthetically, such as by randomly generating optimization problems and applying different algorithms to solve them. In other cases, the meta-dataset may be created by collecting real-world optimization problems and their corresponding solutions.

Once the meta-dataset is created, it can be used to train the meta-learner using techniques such as supervised learning or reinforcement learning. The meta-learner is then able to predict the best optimization algorithm for a new problem, based on the patterns it learned from the training data.

DATA CLEANING

What's Data Cleaning?

Data cleaning is the process of removing, modifying or formatting data that is incorrect, irrelevant or duplicated.

Meta optimization is a type of machine learning technique that involves training a model to optimize the hyperparameters of another model. Data cleaning is a preprocessing step that is often performed before training a machine learning model. It involves identifying and correcting or removing any errors, inconsistencies, or irrelevant information in the data to improve the accuracy and effectiveness of the model.

In the context of meta optimization, data cleaning may involve performing similar preprocessing steps on the data used to train the meta optimizer itself. This can help ensure that the meta optimizer is able to accurately learn and optimize the hyperparameters of the underlying model.

Some common techniques used for data cleaning in meta optimization include:

1.   Removing duplicates and outliers
2.   Handling missing values

3.   Rescaling or normalizing data
4.   Encoding categorical variables
5.   Feature selection or extraction

Overall, data cleaning is an important step in the machine learning pipeline and can have a significant impact on the performance of both the underlying model and the meta optimizer used to optimize its hyperparameters.

When to Clean the Data?

Once we have the dataset ready, we will clean the data. Meta optimization learning models are typically used to optimize the hyperparameters of machine learning models, and data cleaning is an important step in the pre-processing of data for these models.

Data cleaning should be performed before meta optimization because the quality of the input data can significantly impact the results of the hyperparameter optimization. If the input data contains errors, outliers, or missing values, the optimization algorithm may converge to suboptimal hyperparameters, leading to reduced performance of the machine learning model.

Therefore, it is recommended to perform data cleaning before running a meta optimization learning model. Some common techniques for data cleaning include removing duplicates, handling missing values, and removing outliers. Once the data has been cleaned, it can be used to train and validate the machine learning model, and the hyperparameters can be optimized using the meta optimization learning model

Why is Data Cleaning Necessary?

Data Cleaning helps in preparing the data for Exploratory Data Analysis. Data cleaning is necessary for meta optimization learning models because the quality and accuracy of the input data greatly influence the performance of the models. Meta optimization learning models are designed to learn from data, and if the data is noisy, inconsistent, or contains errors, the models may produce unreliable or inaccurate results.

Data cleaning is the process of identifying and correcting or removing errors and inconsistencies in the data. This can involve tasks such as handling missing data, removing outliers, standardizing formats, and resolving conflicts between different data sources. By cleaning the data, the models can better capture the patterns and relationships within the data and produce more accurate predictions and insights.

In the context of meta optimization learning models, data cleaning is particularly important because these models are often used to optimize other machine learning models. Therefore, any errors or inconsistencies in the input data can be amplified and lead to suboptimal results. By ensuring that the input data is accurate and consistent, data cleaning can help to improve the performance of meta optimization learning models and ultimately lead to better results in downstream applications

How to do Data Cleaning?

We use libraries like Pandas, Numpy to do Data Cleaning and apply the following key steps to determine if we need to clean the dataset.

1. Check how many rows and columns are in the dataset.
2. Look for duplicate features by going through the meta info provided.
3. Identify Numerical and Categorical features in the gathered data and check if formatting is required or not.

Formatting can be something like changing data types of the features, correcting the typos or removing the special characters from the data if there are any. If you are working with real time data, then it's recommended to save the cleaned dataset in the cloud databases before the next steps.

EXPLORATORY DATA ANALYSIS (EDA)

What's EDA?

In simple terms, EDA is nothing but understanding and analyzing the data by using various Statistical Measures (like mean, median) and Visualization Techniques(like Univariate Analysis, Bivariate Analysis etc.).

When to Perform EDA?

After the data cleaning stage. Once the data is cleaned, we perform EDA on cleaned data. Explanatory data analysis (EDA) can be performed in the context of meta optimization learning algorithms in several stages of the process, depending on the specific goals and requirements of the task. Some possible examples are:

- Data preprocessing: Before applying any meta optimization learning algorithm, it is often necessary to preprocess the input data to ensure its quality, completeness, and suitability for the task at hand. EDA can help in this stage by identifying outliers, missing values, data inconsistencies, or other anomalies that may affect the performance of the algorithm. Exploratory visualization tools, such as scatter plots, histograms, box plots, or correlation matrices, can be useful for this purpose.

- Feature selection: Meta optimization learning algorithms often require a subset of relevant features or variables from the input data to optimize their performance. EDA can assist in this stage by identifying the most informative features, the correlations between them, or the patterns of variability across different groups or classes of data. This can be done using feature ranking methods, dimensionality reduction techniques, or clustering algorithms, among others.

- Model evaluation: After applying a meta optimization learning algorithm, it is important to assess its performance and interpret the results in a meaningful way. EDA can provide insights into the strengths and weaknesses of the algorithm, the sensitivity of the results to different parameters or settings, or the robustness of the model to different types of data or scenarios. This can be achieved through visualization, statistical testing, or sensitivity analysis, among other techniques.

- Model explanation: In some cases, the goal of meta optimization learning algorithms is not only to optimize the performance of a model but also to understand the underlying mechanisms or factors that contribute to its success or failure. EDA can help in this stage by providing explanations or interpretations of the model's behavior, such as feature importance scores, decision rules, or visualization of decision boundaries. This can facilitate the communication of the results to non-expert users or stakeholders and increase the trust and transparency of the model

Why is EDA Necessary?

Exploratory Data Analysis is considered as the fundamental and crucial step in solving any Machine Learning use case as it helps us to identify trends, or patterns in the data.

Exploratory Data Analysis (EDA) is necessary in Meta Optimization Algorithm for several reasons:

a. Understanding the data: EDA helps in gaining a better understanding of the data used in Meta Optimization Algorithm. It enables the exploration of data patterns, distributions, and relationships between variables. This information can be used to make informed decisions about which optimization algorithm to use and which hyperparameters to tune.

b. Identifying outliers: Outliers are data points that lie far away from the other data points in the dataset. They can significantly affect the performance of optimization algorithms. EDA can help identify outliers, and the Meta Optimization Algorithm can be adjusted accordingly.

c. Feature selection: EDA can help identify the most important features in the dataset. These features can be used in the Meta Optimization Algorithm to select the most appropriate optimization algorithm and hyperparameters.

d. Improving model performance: EDA can reveal data patterns that optimization algorithms may not be able to detect. By using the insights gained from EDA, the Meta Optimization Algorithm can be refined to improve model performance.

e. Avoiding overfitting: EDA can help identify potential areas of overfitting, where the optimization algorithm may be too closely fitting the training data. By identifying these areas, the Meta Optimization Algorithm can be modified to avoid overfitting and improve generalization performance on unseen data.

Overall, EDA is an essential step in Meta Optimization Algorithm because it helps in gaining a better understanding of the data and selecting the appropriate optimization algorithm and hyperparameters, which are critical for achieving good model performance.

How to Perform EDA?

There are Python libraries like Pandas, Numpy, Statsmodels, Matplotlib, Seaborn, Plotly etc, to perform Exploratory Data Analysis.

While doing EDA, some of the basic common questions we ask are:

a. What are the independent and dependent features/labels in the collected data?
b. Is the selected label/dependent feature Categorical or Numerical?
c. Are there any missing values in the features/variables?
d. What are the summary statistics (like mean etc.) for Numerical features?
e. What are the summary statistics (like mode etc.) for Categorical features?
f. Are the features/variables normally distributed or skewed?
g. Are there any outliers in the features/variables?
h. Which independent features are correlated with the dependent feature?
i. Is there any correlation between the independent features? >So, we will try to understand the data by finding answers to the above questions both Visually (by plotting graphs) and Statistically (hypothesis testing like normality tests).

When we are dealing with larger datasets, then it's a bit difficult to get more insights from the data. Hence, at this stage we sometimes use Unsupervised learning techniques like Clustering to identify hidden groups/clusters in the data which thereby helps us in understanding the data more.

FEATURE ENGINEERING

What's Feature Engineering?

A feature refers to a column in a dataset, while engineering can be manipulating, transforming, or constructing, together they're known as Feature Engineering. Simply put, Feature Engineering is nothing but transforming existing features or constructing new features.

Feature engineering is the process of selecting, extracting, and transforming relevant features from raw data to improve the performance of machine learning models. Meta optimization, on the other hand, is a technique for optimizing the hyperparameters of machine learning models to improve their performance.

In feature engineering meta optimization, the goal is to optimize both the feature engineering process and the hyperparameters of the machine learning model simultaneously to achieve the best possible performance. This approach involves using a search algorithm to explore different combinations of feature sets and hyperparameters, and selecting the best combination based on a defined evaluation metric.

This approach can be particularly useful in situations where there are large amounts of data or complex feature spaces, and manual feature selection is difficult or time-consuming. By automating the feature selection and hyperparameter optimization process, this technique can help to improve the accuracy and efficiency of machine learning models.

When to do Feature Engineering?

Feature Engineering is done immediately after Exploratory Data Analysis (EDA), Feature engineering is the process of selecting and transforming input data to create features that can help machine learning models to make more accurate predictions. It is a critical step in the machine learning workflow that can have a significant impact on the performance of the model.

Feature engineering should be done before model training, during the data preprocessing step. The reason is that the quality of the features is crucial to the performance of the model. The better the features are, the better the model can learn from them and make accurate predictions.

In meta-optimization learning models, feature engineering is also important. Meta- learning involves learning to learn, which means that the model needs to be able to adapt to new tasks quickly. Feature engineering can help in this process by creating features that are generalizable across tasks. It's essential to note that feature engineering is an iterative process. You may need to try different transformations and combinations of features to find the best set of features that will help your model perform well on the task at hand. In summary, feature engineering should be done before training the model, and it's crucial for both traditional and meta-optimization learning models.

Why is Feature Engineering Necessary?

Feature Engineering transforms the raw data/features into features which are suitable for machine learning algorithms. This step is necessary because feature engineering further helps in improving machine learning model's performance and accuracy.

1. Algorithm: Algorithms are mathematical procedures applied on a given data.
2. Model: Outcome of a machine learning algorithm is a generalized equation for the given data and this generalized equation is called a model.

In meta optimization learning algorithms, feature engineering is necessary because it helps to extract useful information from the input data and create new features that can improve the performance of the model. Meta optimization learning algorithms are designed to optimize the hyperparameters of machine learning models. These algorithms work by learning from past experiments and using that knowledge to improve the performance of future experiments (Hugo et al, 2023).

Feature engineering involves selecting and transforming input features to make them more informative for the model. This process can include techniques such as dimensionality reduction, normalization, and feature selection. By carefully engineering features, we can help the model to better understand the underlying patterns in the data and make more accurate predictions.

In meta optimization, feature engineering is particularly important because the goal is to find the best hyperparameters for a given model architecture. The performance of the model is highly dependent on the choice of hyperparameters, and by engineering informative features, we can help the algorithm to better identify the optimal hyperparameters((Hugo et al, 2023)). In summary, feature engineering is necessary in meta optimization learning algorithms to help extract useful information from input data, create new features that improve model performance, and enable the algorithm to identify the best hyperparameters for a given model architecture.

How to do Feature Engineering?

We use libraries like Pandas, Numpy, Scikit-learn to do Feature Engineering. Feature Engineering techniques include:

a. Handling Missing Values
b. Handling Skewness
c. Treating Outliers
d. Encoding
e. Handling Imbalanced data
f. Scaling down the features
g. Creating new features from the existing features

Feature engineering is the process of selecting and transforming raw data into a set of features that can be used by a machine learning model to make predictions. In meta-optimization, the goal is to optimize the hyperparameters of a machine learning model. Here are some steps to perform feature engineering for meta-optimization of learning models:

a. Understand the data: Before you start feature engineering, it is important to understand the data you are working with. This includes understanding the data types, distributions, and relationships between the variables.
b. Select features: Choose the most relevant features from your dataset based on your problem statement and domain knowledge. You can use techniques like correlation analysis, feature importance scores, and domain expertise to select features.
c. Transform features: Transform the features you have selected to improve their quality and usefulness. Common feature transformations include normalization, scaling, and one- hot encoding.
d. Create new features: You can create new features by combining existing features or by extracting features from text or images using techniques like word embeddings and convolutional neural networks.
e. Evaluate feature performance: Use a meta-optimization approach to evaluate the performance of your features. This can involve using techniques like grid search, randomized search, or Bayesian optimization to optimize the hyperparameters of your learning model.
f. Iterate: Iterate through steps 2-5 to continuously improve the quality of your features and optimize the performance of your learning model.

Overall, feature engineering for meta-optimization requires a combination of domain knowledge, statistical analysis, and machine learning expertise. It is an iterative process that requires careful evaluation of the performance of your features and learning model.

FEATURE SELECTION

Feature selection is a process of selecting relevant features or variables from a dataset to improve the performance of a machine learning model.

What's Feature Selection?

Feature Selection is the process of selecting the best set of independent features or columns that are required to train a machine learning algorithm.

Feature selection meta-optimization for learning models refers to the process of selecting the most relevant features or variables from a given dataset to be used in building a predictive model. The goal is to find the most informative features that can lead to the best performance of the model in terms of accuracy, precision, recall, and other evaluation metrics.

Meta-optimization involves tuning the hyperparameters of the feature selection algorithms to optimize their performance. This is done by using meta-learning techniques such as cross-validation, grid search, or Bayesian optimization to find the best combination of hyperparameters that lead to the optimal performance of the feature selection algorithm.

There are various methods for feature selection, including filter methods, wrapper methods, and embedded methods. Filter methods select features based on their statistical properties, while wrapper methods use the learning algorithm to evaluate subsets of features. Embedded methods incorporate feature selection within the learning algorithm itself. In summary, feature selection meta-optimization for learning models involves finding the optimal combination of feature selection algorithms and their hyperparameters to select the most informative features from a given dataset for building the best possible predictive model.

There are several approaches to perform feature selection, including filter methods, wrapper methods, and embedded methods. Here is a brief overview of each approach:

a.   Filter methods: Filter methods use statistical measures to evaluate the relevance of each feature in the dataset independently of any particular learning algorithm. Some commonly used statistical measures include correlation coefficients, mutual information, and chi-square tests.

b. Wrapper methods: Wrapper methods evaluate the performance of a learning algorithm using a subset of features. This approach requires a specific learning algorithm to evaluate subsets of features iteratively until an optimal set of features is found. It involves more computational resources than filter methods but can provide better results.

c. Embedded methods: Embedded methods combine feature selection with the learning algorithm. These methods select features during the training process by penalizing or regularizing the model parameters. Examples of embedded methods include Lasso regression and Ridge regression.

The architecture of a feature selection system would depend on the specific approach chosen. For filter methods, the architecture would involve selecting a statistical measure, applying it to the dataset, and selecting the top-ranked features. For wrapper methods, the architecture would involve selecting a learning algorithm, evaluating subsets of features iteratively, and selecting the best performing subset. For embedded methods, the architecture would involve modifying the learning algorithm to include feature selection as part of the training process.

When to do Feature Selection?

Feature Selection is performed right after the feature engineering step. Meta-optimization of learning models involves tuning hyperparameters of a machine learning model to improve its performance. Feature selection can be an important step in meta-optimization, as it can help to identify the most relevant features for a particular model and dataset, and can reduce overfitting.

Feature selection can be done before or during meta-optimization. Before meta-optimization, you can use techniques such as correlation analysis, univariate feature selection, or model-based feature selection to identify the most relevant features for the dataset. This can help to reduce the dimensionality of the data and improve the efficiency of the meta-optimization process. During meta-optimization, you can use techniques such as iterative feature selection, which involves training the model with different subsets of features and selecting the subset that performs best. Another approach is to use regularization techniques such as L1 regularization, which can help to automatically select a subset of the most relevant features while training the model.

The choice of when to perform feature selection depends on the specific problem, the size and complexity of the dataset, and the resources available for meta-optimization. In general, it is recommended to perform feature selection before meta-optimization to reduce the dimensionality of the data and improve the efficiency of the process. However, iterative feature selection and regularization techniques can be useful during meta-optimization to further refine the feature selection and improve the performance of the model

Why is Feature Selection Necessary?

Feature Selection is necessary for the following reasons:

1. Improves Machine Learning Model performance.
2. Reduces training time of machine learning algorithms.
3. Improves the generalization of the model.

How to do Feature Selection?

We use Python libraries like Statsmodels or Scikit-learn to do feature selection. Each of the following methods can be used for selecting the best independent features:

1. Filter methods
2. Wrapper methods
3. Embedded or intrinsic methods

If the number of selected input features are very large (probably greater than the number of rows/records in the dataset), then we can use Unsupervised learning techniques like Dimensionality Reduction at this stage to reduce the total number of inputs to the model.

MODEL ANALYSIS

What's Model Analysis?

Hyper meta learning model analysis involves evaluating the performance of a meta-learning model on a set of related tasks and using this information to optimize the model's hyperparameters. The goal of hyper meta learning model analysis is to improve the model's ability to quickly adapt to new tasks by identifying the most effective set of hyperparameters for the given set of tasks.

To perform hyper meta learning model analysis, the model is first trained on a set of related tasks, and its performance on these tasks is evaluated. The results of this evaluation can then be used to optimize the model's hyperparameters, such as the learning rate, weight decay, and network architecture. This optimization process can be done using techniques such as grid search, random search, Bayesian optimization, or gradient-based optimization.

Once the hyperparameters are optimized, the model is tested on a new set of tasks to evaluate its performance. This evaluation provides feedback on the effectiveness of the hyperparameters and can be used to further refine the model.

Hyper meta learning model analysis is an important step in the development of effective meta- learning algorithms, as it allows models to be optimized for specific sets of tasks and improve their ability to quickly adapt to new tasks. By leveraging prior experience from related tasks and optimizing hyperparameters, hyper meta learning models can improve their performance and generalize to new tasks with higher accuracy.

When Should You Build a Model Analysis?

Hyper meta learning model analysis should be conducted whenever you want to develop a meta- learning model that can quickly adapt to new tasks by leveraging prior experience from related tasks. Hyper meta learning models are especially useful in situations where there is limited training data available or where new tasks are frequently encountered. Here are some situations where building a hyper meta learning model analysis can be particularly beneficial:

a.   Limited training data: When there is limited training data available, hyper meta learning models can be used to learn from related tasks and quickly adapt to new tasks with high accuracy.

b.   Complex and diverse tasks: When tasks are complex and diverse, it can be difficult to train a single model that performs well across all tasks. Hyper meta learning models can be used to learn from related tasks and quickly adapt to new tasks, improving overall performance.

c.   Real-time learning: In some applications, such as robotics and autonomous systems, it is important for models to quickly adapt to new situations in real-time. Hyper meta learning models can be trained on a set of related tasks and then used to quickly adapt to new tasks as they arise.

d.   Continual learning: In situations where new tasks are frequently encountered, it may be necessary to continually adapt the model to new tasks. Hyper meta learning models can be used to quickly adapt to new tasks and improve performance over time.

Hyper meta learning model analysis should be conducted when there is a need to quickly adapt to new tasks and when prior experience from related tasks can be leveraged to improve performance.

Why Should You Build a Model Analysis?

Building a hyper meta learning model analysis can provide several benefits, including:

a.   Improved performance: By optimizing the hyperparameters of a meta-learning model through analysis, you can improve its performance on related tasks and its ability to quickly adapt to new tasks.

b.   Reduced training time: Hyper meta learning model analysis can reduce the amount of time and resources needed to train a model by leveraging prior experience from related tasks and optimizing hyperparameters.

c.   Increased efficiency: Hyper meta learning models are designed to quickly adapt to new tasks, making them more efficient than traditional machine learning models that require a large amount of training data.

d.   Generalization to new tasks: Hyper meta learning models are designed to generalize to new tasks, making them ideal for applications where new tasks are frequently encountered.

e.   Improved scalability: Hyper meta learning models can be scaled to handle large and complex datasets, making them useful in applications where large amounts of data are available.

Building a hyper meta learning model analysis can help to improve the performance, efficiency, and scalability of machine learning models, making them more effective in a wide range of applications.

How Should You Build a Model Analysis?

Building a hyper meta learning model analysis can provide several benefits, including:

a.   Improved performance: By optimizing the hyperparameters of a meta-learning model through analysis, you can improve its performance on related tasks and its ability to quickly adapt to new tasks.

b.   Reduced training time: Hyper meta learning model analysis can reduce the amount of time and resources needed to train a model by leveraging prior experience from related tasks and optimizing hyperparameters.

c.   Increased efficiency: Hyper meta learning models are designed to quickly adapt to new tasks, making them more efficient than traditional machine learning models that require a large amount of training data.

d.   Generalization to new tasks: Hyper meta learning models are designed to generalize to new tasks, making them ideal for applications where new tasks are frequently encountered.

e. Improved scalability: Hyper meta learning models can be scaled to handle large and complex datasets, making them useful in applications where large amounts of data are available.

f. Task selection: The first step is to select a set of related tasks that the model will be trained on. These tasks should be representative of the types of tasks the model will need to adapt to in the future.

g. Model architecture selection: The next step is to select the architecture of the meta- learning model, including the number of layers, the type of activation functions, and the type of optimization algorithm.

h. Hyperparameter optimization: The next step is to optimize the hyperparameters of the meta-learning model using techniques such as grid search, random search, Bayesian optimization, or gradient-based optimization. This involves varying the hyperparameters and evaluating the performance of the model on the set of related tasks to determine the optimal hyperparameter values.

i. Model training: Once the hyperparameters have been optimized, the meta-learning model is trained on the set of related tasks. This involves iteratively adapting the model to each task, using the prior experience gained from the other tasks to improve performance.

j. Model evaluation: After training the model, it is evaluated on a new set of tasks to assess its performance and generalization ability. This evaluation provides feedback on the effectiveness of the hyperparameters and can be used to further refine the model.

k. Deployment: Finally, the model can be deployed in a real-world application, where it can be used to quickly adapt to new tasks by leveraging prior experience from related tasks.

Building a hyper meta learning model analysis involves selecting related tasks, optimizing hyperparameters, training and evaluating the model, and deploying it in a real-world application. By following these steps, you can develop an effective meta-learning model that can quickly adapt to new tasks and improve performance over time.

MODEL BUILDING

What's Model Building?

Building a machine learning model is about coming up with a generalized equation for data using machine learning algorithms.

Machine learning algorithms are not only used to build models but sometimes they are also used for filling missing values, detecting outliers, etc.

When Should You Build a Model?

You start building immediately after feature selection, with independent features.

Why is Model Building Necessary ?

Building a machine learning model helps businesses in predicting the future.

How to Build a Model?

Scikit-learn is used to build machine learning models. Basic Steps to create a machine learning model:

a. Create two variables to store Dependent and Independent Features separately.
b. Split the variable(which stores independent features) into either train, validation, test sets or use Cross validation techniques to split the data.
c. Train set: To train the algorithms
d. Validation set: To optimize the model
e. Test set: To evaluate the model.

Cross validation techniques are used to split the data when you are working with small datasets.

1. Build a model on a training set.
2. What models can you build?

Machine Learning algorithms are broadly categorized into two types, Supervised, Unsupervised machine learning algorithms. Predictive models are built using Supervised Machine Learning Algorithms. The models built using supervised machine learning algorithms are known as **Supervised** Machine Learning Models.

There are two types of Supervised Machine Learning Models that can be build:

1. **Regression models**: Some of the regression models are Linear Regression, Decision Tree Regressor, Random Forest Regressor, Support Vector Regression.
2. **Classification models**: Some of the classification models are Logistic Regression, K- Nearest Neighbors, Decision Tree Classifier, Support Vector Machine(classifier), Random Forest Classifier, XGBoost.

3. **Unsupervised** machine learning algorithms are not used to build models, rather they are used in either identifying hidden groups/clusters in the data or to reduce dimensions of the data. Some of the unsupervised learning algorithms are Clustering Algorithms(like K- means clustering, etc), Dimensionality Reduction Techniques(like PCA etc).

MODEL EVALUATION

What's Model Evaluation?

In simple model evaluation means checking how accurate the model's predictions are, that is determining how well the model is behaving on train and test dataset. Model evaluation is the process of assessing the performance of a machine learning model on a particular task or dataset. In hyperparameter optimization, it is important to evaluate the performance of the model on a validation set during the training process to ensure that the hyperparameters chosen result in a model with good generalization performance.

There are several common metrics used for evaluating the performance of machine learning models, including accuracy, precision, recall, F1 score, and area under the receiver operating characteristic curve (AUC-ROC). The choice of metric depends on the specific problem being solved and the goals of the project.

During hyperparameter optimization, different hyperparameter settings are evaluated by training models on a portion of the available data (the training set) and evaluating their performance on a separate portion of the data (the validation set). This process is repeated for different hyperparameter settings until the best performing set of hyperparameters is found.

It is important to note that model evaluation is an iterative process that requires careful attention to prevent overfitting to the validation set. To address this issue, techniques such as cross- validation and early stopping are often used to prevent overfitting and improve the generalization performance of the model

When to Evaluate the Model?

As soon as model building is done, the next step is to evaluate it. values for the hyperparameters of a given model architecture. Meta-learning refers to the process of learning to learn, which involves the use of machine learning algorithms to learn how to optimize other machine learning models. A hyper-meta-learning model combines these two techniques to learn how to optimize hyperparameters for a given machine learning model. The hyper-meta-learning model learns from previous hyperparameter optimization tasks and uses this knowledge to optimize the hyperparameters for a new task.

The evaluation of a hyper-meta-learning model depends on the specific problem and the performance metric used. In general, the hyper-meta-learning model should be evaluated on its ability to generalize to new tasks and its ability to improve the performance of the machine learning models it optimizes. The evaluation of a hyper-meta-learning model can be done using cross-validation or holdout validation on a set of validation tasks that are separate from the training tasks. The hyperparameters of the meta-learning model itself can be tuned using techniques such as grid search or Bayesian optimization.

The evaluation of a hyper-meta-learning model should be done on its ability to generalize to new tasks and improve the performance of the machine learning models it optimizes, using validation tasks and appropriate performance metrics.

Why is Model Evaluation Necessary?

In general, we will build many machine learning models by using different machine learning algorithms, hence evaluating the model helps in choosing a model which is giving best results. Hyper meta learning models are designed to automate the process of selecting the optimal hyperparameters for a given machine learning model. However, to ensure that the hyperparameters chosen by the model are indeed optimal, it is necessary to evaluate the model's performance on a held-out validation set. Hyper meta learning model evaluation is necessary for several reasons. First, it ensures that the hyperparameters chosen by the model are not overfitting to the training set. Overfitting occurs when a model is too complex and learns to fit the noise in the training data rather than the underlying patterns. If the hyperparameters are chosen based solely on the training set, there is a risk that they will overfit to the training data and perform poorly on new data.

Second, hyper meta learning model evaluation ensures that the chosen hyperparameters are robust and generalize well across different datasets. Different datasets may have different characteristics and require different hyperparameters for optimal performance. By evaluating the model on a held-out validation set, we can ensure that the chosen hyperparameters are not only optimal for the training set but also generalize well to new data ((Théo et al, 2018)).

How to Evaluate a Model?

We use the Scikit-learn library to evaluate models using evaluation metrics. Metrics are divided into two categories as shown:

a. **Regression Model Metrics**: Mean Squared Error, Root Mean Squared Error, Mean Absolute Error.
b. **Classification Model Metrics**: Accuracy (Confusion Matrix), Recall, Precision, F1- Score, Specificity, ROC (Receiver Operator Characteristics), AUC (Area Under Curve).

Evaluating a hyper meta learning optimization model involves assessing its ability to effectively learn and optimize the hyperparameters of other machine learning models. Here are some steps you can follow to evaluate such a model:

a. Define a performance metric: Start by defining a performance metric that will be used to evaluate the performance of the hyper meta learning optimization model. This metric should reflect the quality of the hyperparameters found by the model.

b. Select a dataset: Choose a dataset that is representative of the types of problems the model will be used to optimize. This dataset should have a set of hyperparameters that are known to perform well.

c. Train and validate the model: Split the dataset into training and validation sets. Use the training set to train the hyper meta learning optimization model on a set of machine learning models with known hyperparameters. Use the validation set to evaluate the performance of the hyper meta learning optimization model by comparing the hyperparameters selected by the model with the known optimal hyperparameters.

d. Assess the performance: Calculate the performance metric on the validation set for the hyperparameters selected by the model and compare it to the performance achieved by the known optimal hyperparameters. This will give you an idea of how well the hyper meta learning optimization model is able to learn and optimize the hyperparameters of other machine learning models.

e. Repeat with different datasets: Repeat steps 2 to 4 with different datasets to evaluate the robustness of the hyper meta learning optimization model.

f. Compare with other models: Finally, compare the performance of the hyper meta learning optimization model with other state-of-the-art hyperparameter optimization models to determine how it performs in comparison.

MODEL OPTIMIZATION

What's Model Optimization?

Most of the machine learning models have some hyperparameters which can be tuned or adjusted. For example: Ridge Regression has hyperparameters like regularization term, similarly Decision Tree model has hyperparameters like desired depth or number of leaves in a tree.

The process of tuning these hyperparameters to determine the best combination of hyperparameters to increase model's performance is known as hyperparameter optimization or hyperparameter tuning (Théo et al, 2018).

Model optimization in the context of hyper meta learning algorithms refers to the process of improving the performance of the model used for hyperparameter optimization. In hyper meta learning, a meta-model is trained on a set of machine learning models and their corresponding hyperparameters, and then used to select the optimal hyperparameters for a new model. The meta-model can be optimized in several ways to improve its performance, including:

a. Tuning hyperparameters: The meta-model itself has hyperparameters that can be tuned to improve its performance, such as the number of layers, the learning rate, or the regularization strength.
b. Feature engineering: Feature engineering can be used to extract more informative features from the input data, which can improve the performance of the meta-model. This can involve selecting the most relevant features, transforming the features, or combining them in novel ways.
c. Ensembling: Ensembling involves combining multiple meta-models to improve their performance. This can involve combining different types of meta-models, or training multiple instances of the same type of meta-model with different hyperparameters or feature sets.
d. Transfer learning: Transfer learning can be used to transfer knowledge from one task to another. In the context of hyper meta learning, this can involve using a pre-trained meta- model as a starting point for training a new meta-model.
e. Regularization: Regularization techniques such as dropout or weight decay can be used to prevent overfitting of the meta-model to the training data.

Model optimization is an important part of hyper meta learning algorithms as it can significantly improve their performance and generalization ability

When to Optimize the Model?

After calculating the Evaluation Metrics, we will choose the models with the best results and then tune hyperparameters to enhance the results. Hyperparameter optimization is typically performed after the model architecture has been defined and before the model is trained on the actual data. This is because hyperparameters are tuning parameters that define the behavior of the learning algorithm itself, rather than the model's learned parameters. The optimal time to perform hyperparameter optimization can depend on various factors, such as the size and complexity of the dataset, the computational resources available, and the specific learning algorithm being used. In general, it is recommended to perform hyperparameter optimization early in the development process to ensure that the model is using the best possible hyperparameters. However, it is also important to avoid overfitting the hyperparameters to a particular dataset, which can lead to poor generalization performance on new data. Therefore, it is often useful to perform some initial hyperparameter tuning on a smaller subset of the data, and then perform a final tuning on the full dataset once a promising set of hyperparameters has been identified

Why is Model Optimization Necessary?

Optimization increases the performance of the machine learning models which in turn increases the accuracy of the models and gives best predictions. Hyperparameter optimization is necessary for meta-learning models because the performance of these models depends heavily on the choice of hyperparameters. Meta-learning models are designed to learn how to learn, meaning that they are capable of adapting to new tasks and data by updating their internal representations or parameters (Huaxiu et al, 2021). However, the way they adapt to new tasks and data is determined by their hyperparameters, which need to be optimized for the specific task at hand.

Meta-learning models typically involve a two-level optimization process, where the outer optimization loop updates the meta-parameters (i.e., the hyperparameters), while the inner optimization loop updates the model parameters based on the training data. The hyperparameters control the way the model is updated during the inner optimization loop, and can include parameters such as the learning rate, regularization strength, or the number of layers in the model (Liu et al, 2018).

If the hyperparameters are not properly optimized, the meta-learning model may fail to generalize to new tasks or exhibit poor performance on the training data. Therefore, it is important to perform hyperparameter optimization in order to find the best possible set of hyperparameters for the specific task at hand (Jia et al, 2023). This can involve using techniques such as grid search, random search, or Bayesian optimization to search the hyperparameter space and find the optimal values.

Model optimization involves selecting the best combination of these factors in order to achieve the best possible performance on the task at hand. This can involve a variety of techniques, such as selecting the best algorithm for the task, selecting the best model architecture, tuning hyperparameters, and using techniques such as regularization to prevent overfitting.

Without model optimization, a machine learning model may not perform well on the task it is designed for, or it may suffer from issues such as overfitting or underfitting. Overfitting occurs when a model is too complex and learns to fit the training data too closely, resulting in poor generalization to new data. Underfitting occurs when a model is too simple and is unable to capture the underlying patterns in the data, resulting in poor performance on both the training and test data (Huaxiu et al, 2021).

Therefore, model optimization is necessary in order to ensure that a machine learning model is able to generalize well to new data and achieve the best possible performance on the task at hand.

How to do Model Optimization?

We make use of libraries like Scikit-learn etc or we can use frameworks like Optuna to optimize by tuning hyperparameters (Jia et al, 2023).

Hyperparameter tuning approaches include:

1. Grid Search: In this technique, a set of possible values for each hyperparameter is defined, and all possible combinations of these values are tested on the training data. The combination of hyperparameters that yields the best performance on a validation set is chosen as the optimal set of hyperparameters.
2. Random Search: In this technique, random combinations of hyperparameters are selected and tested on the training data. This approach can be more efficient than grid search for high-dimensional hyperparameter spaces.
3. Bayesian Optimization: This is a more advanced technique that uses Bayesian inference to build a probabilistic model of the objective function (i.e., the model performance) and select hyperparameters that are likely to yield the best performance. This method can be more efficient than grid search or random search for complex optimization problems.
4. Gradient-based optimization: In this technique, the hyperparameters are optimized using gradient descent, where the gradient of the objective function with respect to the hyperparameters is computed and used to update the hyperparameters. This approach can be computationally expensive, but it can be effective for certain types of meta-learning models.

It is important to note that hyperparameter optimization should be performed on a separate validation set, rather than the training set, in order to avoid overfitting to the training data. The best set of hyperparameters should be chosen based on their performance on the validation set, and the final model performance should be evaluated on a separate test set (Liu et al, 2018).

CHALLENGES AND LIMITATIONS OF HYPER META LEARING MODELS

Hyper meta learning models have several challenges and limitations that need to be addressed to improve their performance and scalability. Here are some of the key challenges and limitations:

a. Data scarcity: Hyper meta learning models require a large amount of data to train effectively. However, in many real-world applications, data may be scarce or limited, making it challenging to develop effective models (Théo et al, 2018).
b. Model complexity: Hyper meta learning models can be highly complex, requiring significant computational resources and time to train and optimize. This can make it challenging to develop and deploy these models in real-world applications.
c. Generalization ability: Hyper meta learning models may have limited generalization ability, meaning that they may not perform well on tasks that are significantly different from the tasks they were trained on.
d. Transfer learning limitations: Hyper meta learning models may have limitations in their ability to transfer knowledge between tasks. This may be due to differences in the underlying data distributions, feature representations, or task objectives.
e. Evaluation metrics: There may be challenges in defining appropriate evaluation metrics for hyper meta learning models, as traditional metrics such as accuracy may not be sufficient to capture the model's ability to quickly adapt to new tasks.
f. Interpretability: Hyper meta learning models may be highly complex, making it difficult to interpret how they are making decisions. This can limit their usefulness in applications where interpretability is critical, such as healthcare and finance.

FUTURE SCOPE OF HYPER META LEARNING MODELS

Hyper meta learning models have a promising future as they offer several advantages over traditional machine learning models. hyper meta learning models have a wide range of potential applications in various fields, and their future looks bright as new techniques and algorithms are developed to improve their performance and scalability.

Here are some potential future applications and areas where hyper meta learning models can be useful:

a.   Few-shot and zero-shot learning: Hyper meta learning models can be used for few-shot and zero-shot learning, where the model is trained on a small or no data for a new task, respectively. This can be useful in situations where training data is scarce or where new tasks need to be quickly learned.

b.   Autonomous systems and robotics: Hyper meta learning models can be used to develop intelligent autonomous systems and robotics that can adapt to new situations in real-time. This can be useful in applications such as autonomous vehicles, drones, and manufacturing.

c.   Personalization and recommendation systems: Hyper meta learning models can be used to develop personalized recommendation systems that can quickly adapt to the changing preferences of users. This can be useful in applications such as e-commerce, entertainment, and social media.

d.   Healthcare: Hyper meta learning models can be used in healthcare applications such as disease diagnosis, personalized treatment planning, and drug discovery. By leveraging prior experience from related cases, hyper meta learning models can improve diagnosis accuracy and treatment outcomes.

e.   Cybersecurity: Hyper meta learning models can be used to detect and prevent cyber attacks by quickly adapting to new attack patterns and identifying potential threats.

CONCLUSION

Optimizing hyper meta learning models is a crucial task in the field of machine learning, as these models offer several advantages over traditional machine learning models. In this paper, we have provided an epic overview of the process of optimizing hyper meta learning models, including task selection, model architecture selection, hyperparameter optimization, model training, model evaluation, and deployment. Discussed the benefits of hyper meta learning models and their potential future applications in various fields, such as autonomous systems, personalized recommendation systems, healthcare, and cybersecurity. However, also highlighted the challenges and limitations of hyper meta learning models, such as data scarcity, model complexity, and generalization ability. Future research can focus on developing new techniques and algorithms to overcome these challenges and improve the effectiveness and scalability of hyper meta learning models. By providing a comprehensive overview of optimizing hyper meta learning models, this paper can serve as a valuable resource for researchers and practitioners in the field of machine learning.

REFERENCES

Finn, C., Abbeel, P., & Levine, S. (2022). *Optimizing Learning: A Case Study in Meta-Learning for Computer Vision.* arXiv. (https://arxiv.org/abs/1703.03400)

Huaxiu. (2021). *A Survey on Meta-Learning. 35th Conference on Neural Information Processing Systems (NeurIPS 2021).* arXiv. (https://arxiv.org/abs/1810.03548)

Hugo. (2023). A Tutorial. *International conference on Artificial Intelligence.* Springer.

Kurtosis. (2023). Hyperparameter Optimization in Meta-Learning. *Science direct, 260.*

Liu, T. Z., Yu, J., & Tang, F. (2018, August). J.Deep Meta-learning in Recommendation Systems. *Survey (London, England), 37*(4), 11.

Owen, L. (2022). HyperParameter optimization. O'Riley Media.

Wu, J., Liu, X., & Chen, S. (2023). Hyperparameter optimization through context-based meta-reinforcement learning with task-aware representation. *Knowledge-Based Systems, 260.* https://doi.org/ doi:10.1016/j.knosys.2022.110160

Zhang, Y. (2019). *Efficient Hyperparameter Optimization for Deep Meta-Learning.* arXiv. (https://arxiv.org/abs/2103.09268)

Zhenguo. (2023). Meta-Learning: A Survey. Science direct, 15.

Chapter 4
A Study on Evaluation and Analysis of Edge Detection Operators

Pinaki Pratim Acharjya

(iD) https://orcid.org/0000-0002-0305-2661
Haldia Institute of Technology, India

Subhankar Joardar
Haldia Institute of Technology, India

Santanu Koley
Haldia Institute of Technology, India

Subhabrata Barman

(iD) https://orcid.org/0000-0003-4571-2899
Haldia Institute of Technology, India

ABSTRACT

One of the key stages in both image processing and computer vision is edge detection. For analysis and measurement of several fundamental attributes of an object or set of objects in an image, such as area, perimeter, and form, correct identification of the edges of the objects in the image is crucial. The edge detection operators employed in image processing must therefore be thoroughly understood. In this chapter, fundamental theories and comparative assessments of several edge detection operators are discussed along with a proposed improved contour detection scheme for better performance measurement. The technique has been used to process a number of digital photos, and improved performance in terms of contour detection has been attained.

DOI: 10.4018/978-1-6684-7659-8.ch004

INTRODUCTION

Digital images are a two dimensional light intensity function obtained by a sampling procedure, which is usually referred to as discretization, which transforms the two-dimensional (2-D) continuous spatial signal $f(x,y)$ into the two-dimensional (2-D) discrete, digital image $f(m,n)$. This image represents the response of some sensor (or simply a value of some interest) at a series of fixed positions ($m = 1, 2,..., M;$ $n = 1, 2,..., N$). The indices m and n, respectively, represent the image's rows and columns. The individual picture elements, or pixels, of the image. $f(m, n)$ denotes the response of the pixel located at the *mth* row and *nth* column starting from a top-left picture origin. There are several imaging systems that allow the use of different image sources and the column-row convention. The size of the 2-D pixel grid and the amount of data saved for each individual picture pixel affect the image's spatial resolution and colour quantization. The resolution of an image determines how large or representational it can be. The following three measures can be used to gauge the sharpness of a camera.

Dimensional Resolution

The row column dimensions (m by n) of the image determine the amount of pixels required to fill the visual space that the image captures. The sampling of the visual signal is related to what is commonly known as the pixel or digital resolution of the image. It is commonly referred to by the acronym $m\,n$ (A. Mahmoud, et al., 2016).

Temporal Resolution

This is the total number of images that were collected over a predetermined amount of time in a continuous capture system like video. A typical unit of measurement is frames per second (fps), which describes a single image in a video frame (El Shair, et al., 2022).

Bit Resolution

It relates to the quantization of the image data and indicates the range of intensity/color values that a pixel may have. For example, a binary image only has the two colours black or white. In contrast, a grey-scale image typically has 256 discrete grey levels, ranging from black to white, and a colour image can have any number of hues, depending on the colour space used. In this context, the term "bit resolution" refers to the number of binary bits that are used to store data at a certain quantization level, such as "binary," which is defined as 2 bits, "grayscale," which is defined as 8 bits, and "colour," which is most usually defined as 24 bits. On occasion, a picture's dynamic range is determined by the range of possible values that a pixel can have (S. Ben Aziza, et al., 2015).

Picture element is referred to by the acronym "pixel." It is referred to as the "x, y" or "column-row" distance from the image's origin and represents the smallest, most fundamental part of the image. A digital image with a numerical value is the fundamental informational unit with the specified spatial resolution and quantization. A tiny point sample of coloured light from the scene is often present in each pixel, which determines the image's colour or intensity response. Images occasionally contain information beyond just the visual. A picture is just a 2-D signal that has been digitally transformed into a grid of pixels, some of whose values may be related to characteristics other than colour and brightness.

Morphology is the field of study concerned with form or structure. In order to find and extract useful image descriptors from a picture based on its form or shape properties, we use mathematical morphology. Along with automated counting and inspection, segmentation is one of the key application areas. In the context of set theory, morphology is a significant and potent corpus of techniques that can be precisely mathematically explained. Although this set-theoretic paradigm has the benefits of mathematical rigour, readers without a background in academia may find it challenging to understand. The knowledge of the fundamental concepts and applications of morphology is substantially aided by a practical and intuitive presentation (Sun Zhenlei, 2022).

Binary Images

A binary image only has two different, logical meanings for each pixel: 1 or 0. Foreground, background, and connectivity are the three main components in binary images. In image processing, the foreground and background of a binary image are respectively represented by pixels with logical values of 1 and 0, respectively. Any collection of linked pixels constitutes an object in a binary image. There are two widely accepted meanings of link between pixels. To be regarded as a part of the same object, a foreground pixel needs to have at least one adjoining foreground pixel to its north, south, east, or west. The "4-connection" rule is the name for this requirement (Khalid Aznag, et al., 2020). But when a foreground pixel is sufficiently close to a background pixel in one of the following directions—north, east, north-west, south, or south-west—we use an 8-connection. The following figure serves as an example of these basic ideas .

Segmentation is the division of an image into its separate items or areas. In general, completely autonomous segmentation is one of the most difficult problems to solve when building computer vision systems and is still a hot topic in machine

Figure 1. Pixels in the binary image above are under 4-connected and under 8-connected (groups of connected pixels)

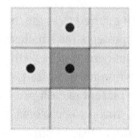

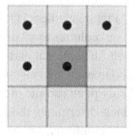

4-connectivity   8-connectivity

learning and image processing. Segmentation is a key component of image processing since it is typically the first and most important step that must be accomplished before further tasks, such as feature extraction, classification, and description, can be attempted logically. After all, how can something be classified or explained if it cannot even be initially identified? As a result, the main goal of segmentation is to separate the image into pieces that are exclusive of one another. Then, we can apply helpful labels. Normally, the foreground and everything else in the image are referred to as the foreground and the background, respectively. One thing to keep in mind is that there isn't one, "correct" segmentation for every single image. Instead, the type of item or region that we are trying to recognise has a big impact on how the image is properly segmented. What kind of relationship must exist between a specific pixel and its surrounding pixels and other pixels in the image for it to be categorised as being in one area or another? In fact, this is the essential question in picture segmentation, and there are generally two ways to answer it:

Methods for Detecting Edges and Boundaries

Images, which are defined as a large local shift in the intensity of the image, typically include discontinuities in their intensity or first derivative around their edges. There are two different types of image intensity discontinuities: (1) step discontinuities, where the intensity of the image abruptly changes from one value on one side of the split to a different value on the opposite side and (2) Line edge, where the image intensity abruptly changes values and quickly returns to the initial value (Javier E. Santos, et al., 2022). But in actual images, step and line limitations are rarely visible. Real signals rarely feature sharp discontinuities because of the existence of low-frequency components or the smoothing effect that is introduced in the image during image acquisition. Step edges serve as the foundation for ramp and line edges.

Edge detection is the process of identifying significant local discontinuities in an image. A step edge in one dimension corresponds to a local peak in the first derivative. An image can be conceptualised as an array of pixel intensities of a continuous function of image intensity values. The gradient is a measurement of change in a function. This method of establishing the boundary between geographic areas is based on the identification of edges. It looks for definite distinctions between pixel groupings as a consequence. Algorithms for edge detection go through four steps.

Filtering

Filtering is routinely used to improve the effectiveness of an edge detector..

Enhancement

In order to facilitate the detection of edges, it is critical to recognise intensity variations in a point's surrounds. Enhancement, which is frequently done by calculating the gradient magnitude, draws attention to pixels where there has been a noticeable change in the local intensity values.

Detection

Many points in an image have gradient values that are nonzero, even though not all of these points in a specific application are edges. To ascertain whether points are edge points, a procedure must be used. Through thresholding, the detection criterion is provided. Localization: The location of the edge can be calculated at the subpixel level.

You can also guess how the edge is oriented. It is important to remember that detection does not necessarily give a perfect estimate of edge location or orientation; rather, it only indicates the presence of an edge close to a pixel in a picture. The most common edge detection errors are inaccurate and missing edges. These errors fall under the category of classification errors.

Methods Based on Regions

According on the degree of reciprocal similarity, this approach assigns pixels to a specific zone. Image segmentation is the division of a picture into various stages while monitoring key characteristics of each phase. Every image has specific properties, such as colour and objects that are present in the image, and image segmentation is a process for dividing images based on such characteristics. As each of the many phases of the image can be treated differently following a segmentation process, this can be used for both image analysis and additional image processing. A fundamental challenge in computer vision and image processing is edge detection. That has been a very worrying problem for the researchers and image segmentation.

A shift in intensity from one pixel to the next can significantly alter the quality and ability to segment a image in digital photographs, which frequently exhibit dramatic contrasts in intensity along their edges. Image segmentation is the process of dividing an image into sections that are relevant to a given application. To understand an image, computer vision and image processing systems must be able to recognize the borders of each object in the image. For image segmentation and object boundary extraction in digital images, there are numerous edge detection operators available. Each operator is built with a specific kind of edges sensitivity in mind. As significant operators, they have Sobel, Roberts, Prewitt, LoG, and Canny.

Even though many approaches have been put forth over the past few decades, the engineers are still facing a significant challenge with automatic image segmentation. For real-world images to be ready for machine processing, edge detection is a significant issue. It is a crucial area in image processing and picture segmentation. Digital images' edges often have stark intensity contrasts, and a change in intensity from one pixel to the next can significantly change the quality of the image. Due to these factors, edges serve as both the outline of an object and a dividing line between objects that overlap.

There are numerous detection algorithms that can be used for edge and image segmentation tasks. A well-known component is one of them. It has been noted that using the conventional 5x5 mask of the Laplacian and Gaussian edge detector for picture segmentation and edge detection does not also address the primary issue of reliable edge identification. One of the essential building elements in image processing is the gradient of the image. In image processing, the gradient is the desired first-order derivative. A 2D vector at each image point mathematically represents the gradient of a two-variable function, in this case the picture intensity function, with the components marked by the derivatives in the horizontal and vertical axes (A. Sathesh, et al., 2021).

This chapter proposes a modified method of creating gradient images using the Laplacian of Gaussian operator and a 9x9 mask, which results in a higher level of edge detection accuracy. In addition to Peak Signal to Noise Ratio (PSNR), Mean Square Error, and other statistical measures of randomness, entropy can be employed to describe the texture of the output images (MSE). According to the following, this work is organised: Brief information on image segmentation is included in Section 2. In this chapter's section 3, we examined applications for image processing, particularly picture segmentation. When segmenting images, edge detection is a notion that is explained in Section 4. Software called MATLAB is used to carry out procedures for digital images.

IMAGE SEGMENTATION

The process of segmenting an image involves breaking it up into groups or areas that represent various things or components of objects. A number of these categories are assigned to each pixel in an image. Typically, a good segmentation has pixels that belong to the same category that have similar multivariate grey scale values and form a connected region, while nearby pixels from different categories have dissimilar values. The most important step in image analysis is often segmentation because it is at this stage that we stop treating each pixel as a separate unit of observation and start working with objects (or parts of objects) made up of many pixels (L. S. Davis,et al.,1975).

Segmenting images is the key to understanding them. It is believed that image segmentation is a crucial procedure that must be carried out in order to study and interpret obtained images effectively. It is one of the most challenging image processing tasks, but it is also a crucial and essential part of an image analysis and/ or pattern recognition system that affects how effectively the final segmentation functions (Z Wu, et al., 1993. The processes of picture analysis that follow are easier if segmentation is done correctly. However, as we shall see, automated segmentation algorithms frequently only have some degree of success. Even though some issues may occasionally need to be fixed manually, the computer should have finished most of the process by this point. There are three main segmentation techniques:

- Termed thresholding
- Edge-based methods
- Region-based methods.

According to the range of values that they fall inside, pixels are categorised into groups during thresholding. Through thresholding, the image of the muscle fibres was transformed into the limits in Fig. 2(a). Two categories have been created out of the remaining pixels, one for those with values under 128 and the other for all other pixels (T. S. D. Murthy, et al., 2014). White lines that mark the boundaries between neighbouring pixels in several categories are superimposed on the original image. The threshold clearly succeeded in dividing the image into the two primary fibre types.

Applying an edge filter on the image, classifying the pixels as edge- or non-edge based on the filter output, and assigning the pixels that are not divided by edges to the same group are all done through edge-based segmentation (Luxit Kapoor, et al., 2017). The boundaries of connected regions are displayed in Fig. 2(b) after Prewitt's filter has been applied and all non-border segments with less than 500 pixels have been eliminated.

Finally, region-based segmentation algorithms function iteratively by dividing groups of pixels with different values from neighbours and grouping pixels with similar values together. One such method, which is based on the idea of watersheds, is shown in Figure with the boundaries it generated.

It should be noted that none of the three methods shown in Fig. 2(c) has been totally successful in separating every adjacent pair of fibres in the muscle fibre image. Each technique has specific drawbacks. For instance, while certain boundaries are present in Fig. 2(a), others are absent. However, there are more borders in Fig. 2(b), and while they are smooth, they are not always in their proper locations.

The main focus of computer vision research is segmentation. There are several picture segmentation techniques available, but no one methodology is ideal for all

Figure 2. The boundaries were created using three different segmentation techniques: (a) thresholding, (b) connected regions after removing small regions and thresholding the output of the Prewitts edge filter, and (c) the watershed algorithm's result on the output of a variance filter with Gaussian weights. (σ2=96)
[Source: B.P. Nguyen, et al., 2016.]

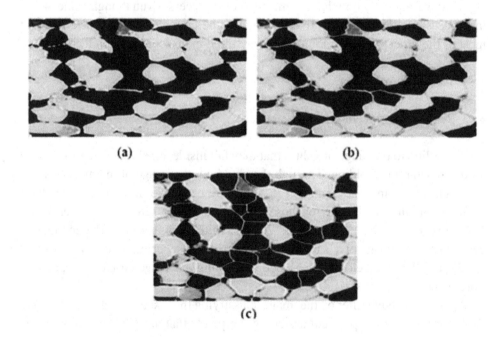

(a)

(b)

(c)

applications. This fundamental issue has been thoroughly studied by researchers, who have also suggested a number of image segmentation approaches (Priyanka Kamra, et al., 2015).

APPLICATIONS OF IMAGE SEGMENTATION

Image segmentation produces either a group of segments that collectively cover the entire image or a series of contours that are taken from the image. Regarding many characteristics or computed properties, such as colour, intensity, or texture, each pixel in a region is comparable. With regard to the shared traits, neighbouring areas diverge significantly. Following are a few examples of how image segmentation is used in practise:

Medical Imaging

Medical abnormalities can be found in clinical scans like CT or MRI scans, and machine learning picture segmentation is essential in this process. Medical professionals can manage their time with the use of quick and precise computer vision algorithms.

It is not necessary to review every scan. By relying on computer vision specialists, medical procedures might be simplified and examination capacity increased (M. Moghbel, et al., 2016).

Locates Tumors and Other Pathologies

The following steps are used to separate the regions of the infected brain using MR imaging: The preprocessed brain MR picture is transformed into a binary image in the first stage, with a cut-off threshold of 128 chosen. Due to these two, several zones are generated around the contaminated tumour tissues, which are clipped out. Pixel values larger than the chosen threshold are mapped to white, while others are indicated as black. In the second step, a morphological erosion process is used to remove white pixels. The original picture and the eroded region are then split into two equal parts, and the black pixel region that was recovered from the erode procedure is tallied as a brain MR image mask (Chao-Lun Kuo, et al., 2017).

Measure Tissue Volumes

The new technique for measuring epicardial fat volume uses threshold-based 3-dimensional (3D) segmentation. A: The initial 3D volumetric data obtained extrathoracic fat tissue in addition to intrathoracic fat (CT attenuation range between

Figure 3. The task of segmenting objects of interest in a medical image, such as teeth, is known as medical image segmentation (orthopantomogram).
[Source: N. V. Blanco, 2019.]

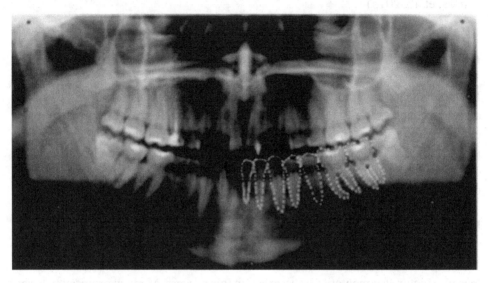

Figure 4. MRI of a sample HGG patient with three major planes of three-dimensional segmentation of various intra-tumoral features (ED, ET, and NCR/NET) (axial, sagittal, and coronal).
[Source: Subhashis Banerjee, et al., 2020.]

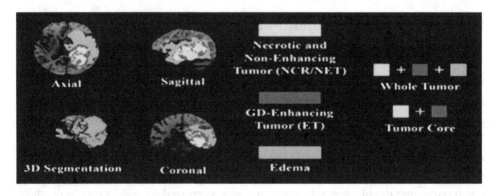

-200 HU and -30 HU). B: It is compared to the appropriate 2-dimensional axial picture to determine the precise volume of epicardial fat. It illustrates 3D epicardial fat volume in surface-shaded display image C. (gray color), Hounsfield unit, or HU (Z. Akkus, et al., 2017).

Figure 5. Measurement of epicardial fat volume with 3-dimensional (3D) segmentation based on thresholds
[Source: Mi Jung Park, 2010]

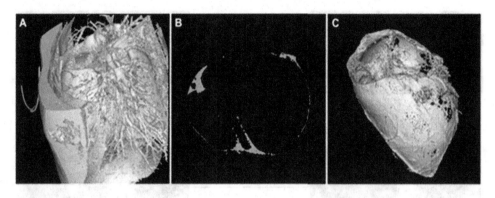

Computer-Guided Surgery

With the aid of images and computer technologies, image guided surgery (IGS) seeks to improve minimally invasive surgery. Preoperative pictures are used in IGS to design the operation. The surgeon is directed throughout the process via intraoperative pictures and instrument tracking. Image segmentation is a crucial stage of the IGS image processing process. By removing extraneous information from the photos, segmentation makes it possible to visualise the structures of interest. For registration and structure analysis, such as determining a tumor's volume, segmentation is also helpful. To be effective in a clinical context, segmentation of images acquired both before and during the surgery must be quick and precise (T. Cootes, et. al., 1999).

Diagnosis

Using a localized deep learning method called You-Only-Look-Once (YOLO), breast mass can be detected from full mammograms (Nitesh Sharma, et al., 2014). Full resolution convolutional network (FrCN), a new deep network model, is suggested and used to segment the mass. Finally, the mass is identified and classified as benign or malignant using a deep convolutional neural network (CNN). The publicly accessible and annotated INbreast database was used to assess the accuracy of detection, segmentation, and classification of the proposed integrated CAD system.

Figure 6. Real-time deep learning semantic segmentation for 3D augmented reality help during intraoperative surgery
[Source: Tanzi, et al., 2021]

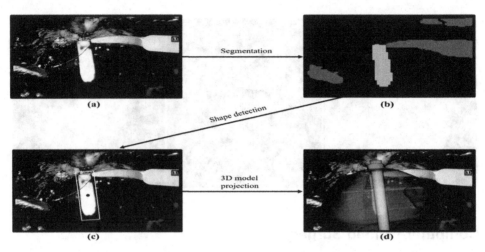

Figure 7. Diagrammatic representation of the proposed deep learning-based computer-aided diagnosis (CAD) system for the detection, segmentation, and classification of breast cancer masses from input digital X-ray mammograms
[Source: Mugahed A. Al-antari, 2018.]

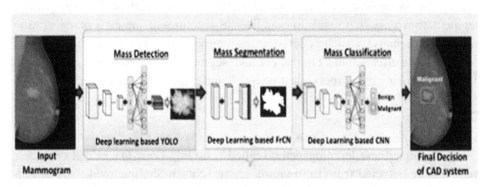

Treatment Planning

The purpose of radiotherapy healing setting up is to support an efficient course of action that will provide a precise dose of radiation to the target volume without harming the nearby healthy tissues. As a result, when designing the irradiation procedure, patient positioning, objective volume specification, as well as irradiation field placement are extremely important steps.

The Beam's Eye View (BEV) (Yaxi Liu, et al., 2009) image, which is produced from the viewpoint of the beam source and is crucial for RTP imaging, may be seen in figure below together with the BEV view direction. The interactive DRR images are shown in the system's equivalent window, known as the BEV window. The doctors can examine the patient anatomy in the DRR image as they study it on the Simulator monitor through the BEV window.

The Observer's Eye View (OEV) window depicts the patient commencing the perspective of the room, simulating the way a doctor would observe a patient in a real-world setting. The graphic below shows the OEV view orientation. The patient's CT scan is used to generate each image in both the BEV and OEV windows, while publicized in the following stature.

Study of Anatomical Structure

The significance of image segmentation algorithms has expanded due to the rapid advancements in imaging modalities. To differentiate between the component of interest and other components, image segmentation separates an image into sections depending on the image's internal components. This procedure examines the component's region of interest in order to gather more precise quantitative information about the component. However, practically all methods for segmenting

Figure 8. The system's six-window design: The OEV is in the centre bottom window on the left, room view is in the lower right, and BEV is in the top right
[Source: S. Zimeras, 2012.]

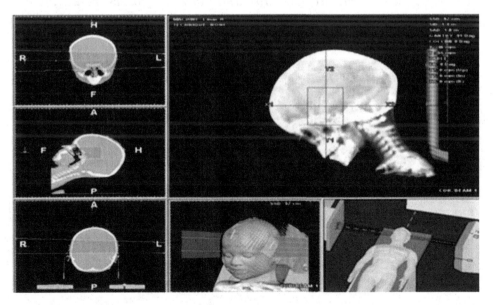

anatomical structures have inherent issues such the necessity for sophisticated parameter adjustments and the existence of picture artefacts (Soon-ja Yeom, 2011).

Locate Objects in Satellite Images

Two strategies are being considered in this situation. Creating tiled versions of the photos is one method (e.g. 64x64 pixels). Use a CNN to identify the parking lot pixels in each tile based on the training data (S. Deepthi, 2021). In order to create superpixel representations of the items in the imagery (such as buildings, roads, and parking lots), the second approach segments the objects in the imagery. It makes use of Felzenzwalb segmentation from SKImage. A random forest method was used to analyse the properties of the superpixels as a backup model (A. Tahir, et al., 2022).

Face recognition

A facial recognition system is a piece of technology that can compare a human face in a digital photo or video frame to a database of faces. Such a technology locates and measures face features from an image and is often used to authenticate individuals through ID verification services (KH Teoh, et al., 2021). Particularly most significant ladder in the face recognition progression is the localization of

Figure 9. Anatomical structure segmentation in 3D medical imaging: (a) Noisy input head-related images from a patient's computed tomography. (b, c, d,) 3D produced views that are segmented and colored. To aid in medical diagnostics, the area around the teeth and the remainder of the skull are given various colors and opacities. [Source: S. Zimeras, 2010]

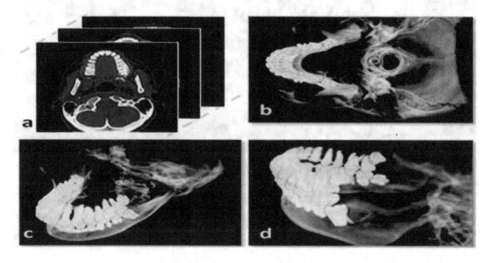

Figure 10. Locating items in satellite pictures using object-based image analysis
[Source: Huge FastAi fan, 2018.]

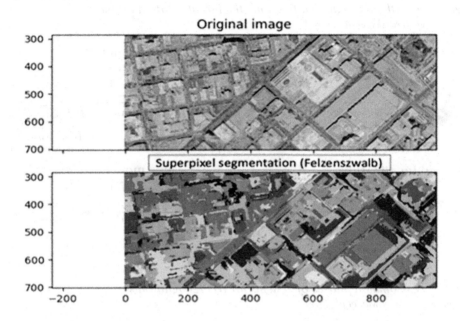

human faces in digital photographs. Here, a shape comparison method is used to quickly and accurately detect faces in changing lighting and backgrounds. A two-step method that enables both rough face localization and detection is provided. On a large test set base, experiments were run and evaluated using a new validation metric (Yan Sun, et al., 2022).

Iris Recognition

Iris recognition has been widely employed in several situations with excellent outcomes. As one of the initial steps, picture segmentation is fundamental to the procedure and is essential to the accomplishment of the recognition task (Alice Nithya, et al., 2015).

Fingerprint Recognition

A person is identified by their fingerprint using a biometric recognition system called the fingerprint recognition system. This system is employed for purposes of identification and verification. By placing a fingertip on the scanner, the user whose impression needs to be identified or validated has their fingerprint registered. The scanner creates

Figure 11. The segmentation and localization stages are contained in the two phases of the face detection system (AOI = area of interest). With a face model, do coarse detection; with an eye model, fine-tune the initial position estimate.
[Source: O. Jesorsky, et al., 2001.]

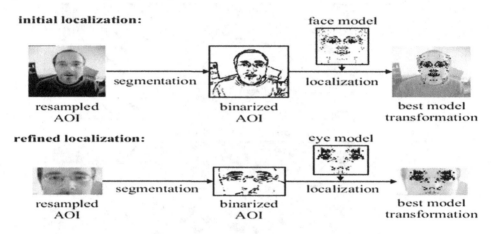

Figure 12. Common phases of systems for iris recognition system
[Source: Sunil Chawla, et al., 2011.]

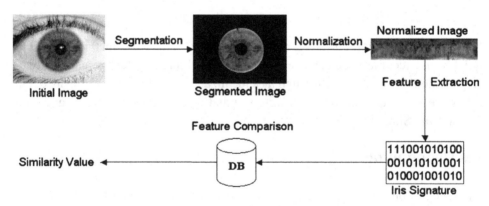

a digital copy of the fingerprint, which is the impression of the fingertip, and compares it to a database template to identify and confirm the person (J. Priesnitz, et al., 2021).

Traffic Control Systems

An efficient management system is required since key crossroads frequently experience traffic congestion. It is crucial to implement a smart traffic controller that uses real-time visual processing. Various item counting techniques and edge identification algorithms are used to evaluate the camera's sequence. Previously, they employed

Figure 13. Claim of identification is made by a module that communicates with the application system and the user
[Source: N. Shanmathi, et al., 2018.]

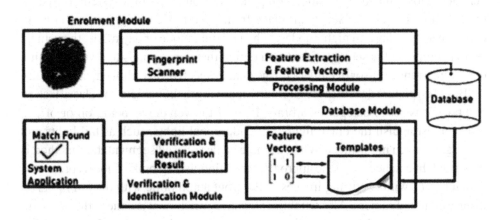

the matching approach, which calls for the installation of the camera alongside the traffic signal. It will record the series of images. A filtering technique is used to clean the image and remove all distracting elements to reveal only the cars, after which it clearly displayed the number of cars in the image, in order to create a picture of a deserted road that can be used as an allusion image. It makes use of software for taking pictures or videos. It has been modified so that it can be used in the future to regulate traffic signal signs by allowing enough time for each sign, susceptible on the quantity of vehicles travelling in both directions (Anand Shah, et al., 2020).

Brake Light Detection

In recent years, front car brake light detection has grown in importance in terms of transportation system safety. As a component of unintentional braking or a cautionary

Figure 14. Traffic control system using video
[Source: Arinaldi, et al., 2018.]

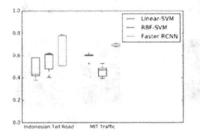

scheme, the identification and bigotry of the brake lights using vehicle-mounted cameras provides early caution to prevent rear-end collisions for the cars. Mode, vehicle region, and categorization components are all parts of a system. Based on the brightness of each pixel in an image frame, the mode component is set up to choose between day and night modes. When in night mode or day mode, the vehicle region component is set up to identify a region that corresponds to a vehicle using data from a range sensor or camera image data, respectively (Pirhonen, et al., 2022).

Based on picture data in the region corresponding to the vehicle, the classification component is set to categories a brake light of the vehicle as being on or off. In order to distinguish the brake lights from additional lights, such as tail lights and turn lights, when nighttime driving, a single camera-based segmentation method is presented. In essence, a unique approach for distinguishing brake lights is proposed, which begins with capturing images of the front automobile with the tail lights on using a mounted camera. After attainment, image enrichment is done to the frames to lighten the rest of the image, counting the centre of the light sources, while whitening the red corona. The calculation of the white and black pixel ratios in coronas uses the collective contrast variations as well as the back light arrangements to identify the region of interest. However, while the ratios of the tail lights and brake lights are roughly the same at all distances, the brake light ratios are noticeably higher, making it possible for nighttime driving vehicles to distinguish between the two types of lights (G. Kumar, et al., 2021).

Figure 15. Brake light detection with front camera of a car and process to detect
[Source: Chen, Hua-Tsung, et. al., 2016.]

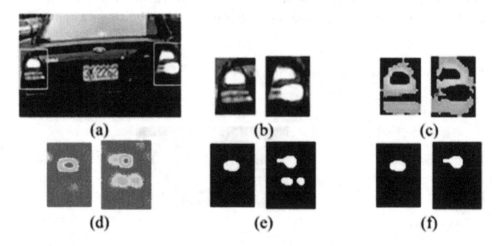

Crop Disease Detection

The main challenge facing farmars is how to assess diseases of diseased leaves. The answer might involve taking a picture of the diseased leaves, determining what caused the infection, and figuring out a cure. The automatic leaf disease diagnosis system for precision agriculture makes use of picture acquisition, processing, segmentation, feature extraction, and machine learning methods. A quick and precise diagnosis of the plant illness is provided to the farmer via an automated disease detection system. To speed up crop diagnosis, plant leaf disease detection systems must be automated. The frameworks are made for spotting leaf sickness using machine learning and image processing (Vijai Singh, et al., 2020).

This framework can take as input a picture of a leaf. To begin with, noise is removed from leaf pictures during preprocessing. To eliminate ambiance noise, use the mean filter. The image quality is improved via histogram equalization. In photography, segmentation is the split of a single picture into a lot of parts or segments. It helps to define the limits of the image. The K-Means technique is employed to segment the picture. The main component analysis is used to carry out feature extraction. The next step is to categories the photos using methods like RBF-SVM, SVM, random forest, and ID3.

Figure 16. Agricultural image collection to disease detection technique
[Source: M.G. Selvaraj, et al., 2019.]

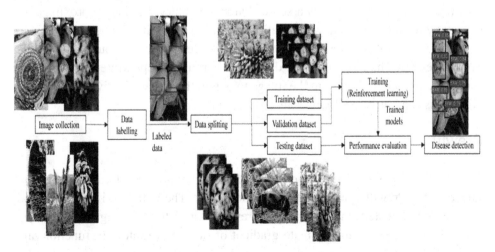

EDGE DETECTION IN IMAGE SEGMENTATION

Edge detection is a phrase that is crucial to both computer vision and image processing. The picture's edges depict the substance's regions. As edges break with difficult intensity contrasts or jump in intensity from one pixel to the next, edges are the main problem in image processing (R. Muthukrishnan, et. al., 2011).

Edge identify picture significantly reduces the amount of data, eliminates useless information, and preserves the important structural assets. It is used by digital artists to create image outlines.

To enhance the edges of an image, the output of an edge detector can be added back. The process of segmenting images frequently starts with this. In below examples of edge detection with three digital images are illustrated where the Figure 17(a), Figure 17(c) and Figure 17(e) are the original images of Coins, cameraman and Lena respectively. The resultant segmented images where the edges are very prominent are shown in Figure 17(b), Figure 17(d) and Figure 17(f) for Coins, cameraman and Lena respectively. The main goal of edge detection is to extract a "line drawing" representation from an image useful for object recognition.

TRADITIONAL EDGE DETECTORS

Sobel

The discrete differences between the rows and columns of a 3X3 neighbourhood are used by the sobel edge detector to calculate the gradient. The Sobel operator is relatively low-cost in terms of calculations because it works by convolution the image with a small, separable, integer-valued filter in both the horizontal and vertical directions. However, it creates a somewhat rudimentary gradient approximation, especially for high frequency fluctuations in the image. An "Isotropic 3x3 Image Gradient Operator" is what it is. Central differences are also a foundation of the Sobel operator. The first Gaussian derivative is roughly represented by this. This is equal to the 33 mask used to the image to obtain the first derivative of the image after Gaussian blurring (P. Ganesan, et al., 2017).

Prewitt

One of the earliest and most well-researched techniques for detecting edges in images is the Prewitt operator edge detection mask. The initial derivatives Gx and Gy are approximated digitally by the Prewitt edge detector using the following mask. It computes an approximate gradient of the picture intensity function and

Figure 17. Examples of edge detection (a) original image of coins, (b) segmented image of coins, (c) original image of cameraman, (d) segmented image of cameraman, (e) original image of Lena, (e) segmented image of Lena
[Source: Debelee, et al., 2019, and Safia, et al., 2016.]

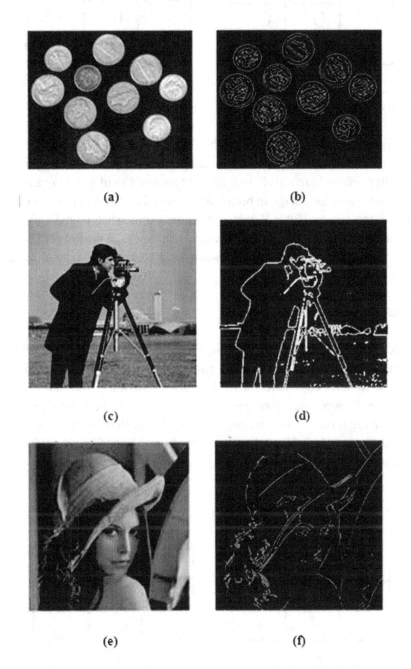

Figure 18. Sobel edge detector

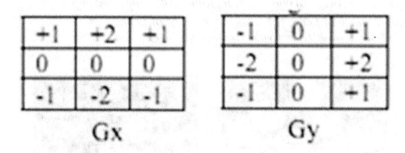

is a discrete differentiation operator. The output of the Prewitt operator at each location in the image is either the associated gradient vector or this vector's norm. The Prewitt operator is relatively low-cost in terms of calculations because it only requires convolving the image in both the horizontal and vertical axes with a tiny, separable, integer-valued filter. However, it creates a somewhat rudimentary gradient approximation, especially for high frequency fluctuations in the image. The Prewitt approach uses the centre difference between the pixels in the surrounding area (P. Selvakumar et al., 2016).

Roberts

The vertical and horizontal edges are brought out separately in Robert edge detection before being combined to produce the edge detection. The following masks are used by the Roberts edge detector to digitally approximate the first derivatives as differences between neighbouring pixels. Lawrence Roberts first suggested it in 1963, making it one of the first edge detectors. The Roberts cross operator is a differential operator with the goal of approximating an image's gradient by discrete differentiation. This is done by computing the sum of the squares of the differences

Figure 19. Prewittl edge detector

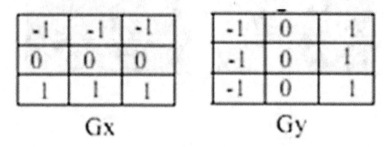

between pixels that are diagonally adjacent. The diagonal elements are the derivatives of Roberts kernels. They are hence referred to as cross-gradient operators. They are based on the disparities in the cross diagonals (Guowei Yang, et al., 2011).

Laplacian of Gaussian (LOG)

This detector searches for zero crossings after applying a Laplacian or Gaussian filter to f(x, y). In this technique, the image is broken up where the intensity varies to efficiently detect the edges using a combination of Laplacian and Gaussian filtering. It determines where the boundaries should be and tests a larger area surrounding the pixel (Tsung-Han Lee, et al., 2022).

When the edge magnitude exceeds the threshold value, edges are said to be present in the first derivative. The edge pixel is present in the second derivative scenario where the second derivative is zero. This is comparable to saying that the sign of the pixel differences changes when the second derivative of f (x) crosses a zero-crossing. One such zero-crossing approach is the Laplacian algorithm. However, there are numerous issues with zero-crossing algorithms. The issue with Laplacian masks is that there is no magnitude testing, so even a tiny ripple can cause the approach to produce an edge point. As a result, the image must be filtered before the edge detection algorithm is used. Although one-pixel thick edges are typically favored, this method results in edges that are two pixels thick. The benefit is that the edge thinning procedure is unnecessary because the zero-crossings themselves identify where the edge points are (Weibin Wu, 2016).

The Laplacian of Gaussian operator is frequently preferred in order to reduce the Laplacian operator's noise sensitivity. The supplied image is first blurred with the Gaussian operator before being processed with the Laplacian operator. The Laplacian minimises the identification of spurious edges since the Gaussian function lowers noise.

Figure 20. Roberts edge detector

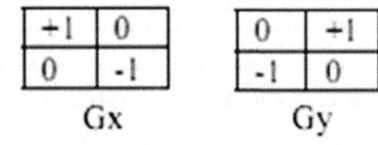

THE GRADIENT

Digital images' edges often have stark intensity contrasts, and a change in intensity from one pixel to the next can significantly alter the quality of the image. Such discontinuities are found using first and second order derivatives. The gradient is the preferred first-order derivative in image processing. A 2-D function's gradient, f(x,y), is represented by the vector

$$\nabla f = \begin{bmatrix} g_x \\ g_y \end{bmatrix} = \begin{bmatrix} \dfrac{\partial f}{\partial x} \\ \dfrac{\partial f}{\partial y} \end{bmatrix}$$

This vector's magnitude is:

$$\nabla f = mag(\nabla f = \left[g_x^2 + g_y^2 \right]^{\frac{1}{2}} = \left[\left(\frac{\partial f}{\partial x} \right)^2 + \left(\frac{\partial f}{\partial y} \right)^2 \right]^{\frac{1}{2}}$$

Sometimes, this amount is approximated by skipping the square root operation,

$$\nabla f \approx g_x^2 + g_y^2$$

the use of absolute values,

$$\nabla f \approx \left| g_x^2 \right| + \left| g_y^2 \right|$$

These approximations nevertheless act as derivatives; in contrast-intense areas, they are zero, and in areas with varying intensity, their values depend on how much the intensity has changed. The term "gradients" is frequently used to refer to the gradient's magnitude or its approximations. In the process of segmenting images, the Laplacian of Gaussian filter (LOG) (Weibin Wu, (2016) is crucial. It is a convolution filter that is employed in the detection of various objects' edges. The Laplacian filter is applied first, followed by a Gaussian blur, and then a check for zero crossings, or when the resultant value changes from positive to negative or from negative to positive. This filter's primary goal is to draw attention to the borders of various things. The LOG operator accepts a single grayscale image as input and outputs a second binary image.

Figure 23 below shows the flowchart for the method for creating gradient images. A coloured image is initially selected and entered into the Mat Lab software for processing. The first step is to convert the image to grey scale. In a grayscale image, black and white are primarily used as combinations of two colours. Black represents the lowest or weakest intensity, and white represents the highest or strongest intensity. The edges of an object or objects are formed by variations in intensity levels. In the last stage, various edge detection operators are used to find the boundaries and gradients of the objects.

Segmentation results with standard/conventional edge detection operators for generating gradient image are shown in below images where in Figure 22(a) original grayscale image of a bird is shown as input, Figure 22(b) is the resultant gradient image with Sobel operator, Figure 22(c) is the resultant gradient image with Prewitt operator, Figure 22(d) is the resultant gradient image with Roberts operator and Figure 22(e) is the resultant gradient image with Laplacian of Gaussian (LoG) operator.

PROPOSED SCHEME

After applying a Laplacian of Gaussian filter on f(x, y), the Laplacian of Gaussian detector searches for zero crossings to identify edges. In this technique, the image is broken up where the intensity varies to efficiently detect the edges using a combination of Laplacian and Gaussian filtering. It determines where the boundaries should be and tests a larger area surrounding the pixel. A standard 5x5 Laplacian of Gaussian edge detection mask is shown below.

A modified 9x9 mask of the Laplacian of the Gaussian operator is offered in the suggested approach. The size of the ideal mask was discovered to be 9x9 in dimensions after numerous attempts with masks with bigger dimensions. This image displays the altered mask.

The suggested scheme's flowchart is shown in the section below. A colour image is initially transformed into a grayscale or black and white image. With the assistance of the suggested modified 9x9 mask of the Laplacian of Gaussian edge detection operator, the gradient picture is gathered from the grayscale image.

Figure 21. Flowchart of the approach of generating gradient images

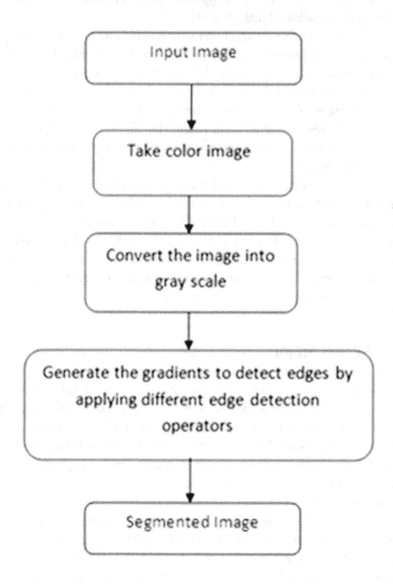

EXPERIMENTAL RESULTS

Two photos taken in the real world Lena's and Bird's respective Figures 26(a) and 26(b) have been chosen for experimental purposes. Figures 27(a) and 27(b) respectively show the resultant gradient images acquired using the traditional 5x5 LoG mask, while Figures 28(a) and 28(b) respectively show

Figure 22. Generating gradient images with conventional edge detection operators

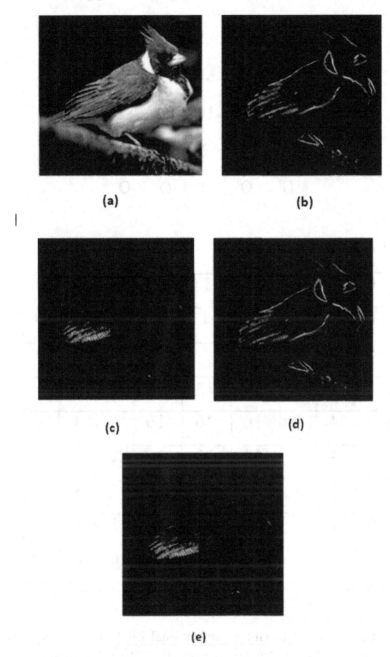

the subsequent gradient images obtained using the suggested 9x9 LoG mask. The statistical analysis results are presented in Table 1 below. The edges of

Figure 23. Conventional 5x5 LoG operator

0	0	-1	0	0
0	-1	-2	-1	0
-1	-2	16	-2	-1
0	-1	-2	-1	0
0	0	-1	0	0

Figure 24. Proposed 9x9 LoG operator

0	0	0	1	1	1	0	0	0
0	1	3	4	4	4	3	1	0
0	3	4	2	0	2	4	3	0
1	4	2	-10	-13	-10	2	4	1
1	4	0	-10	-30	-10	0	4	1
1	4	2	-10	-13	-10	2	4	1
0	3	4	2	0	2	4	3	0
0	1	3	4	4	4	3	1	0
0	0	0	1	1	1	0	0	0

gradient images produced with a conventional Laplacian of Gaussian edge detector of a 5x5 mask are frequently splotchy and unconnected. A modified Laplacian of a Gaussian edge detector with a 9x9 mask, on the other hand, yields significantly more satisfactory results in terms of sharper edges when applied to gradient images. PSNR and MSE for statistical analysis in relation to

Figure 25. Flowchart of generating gradient images with proposed 9x9 LoG

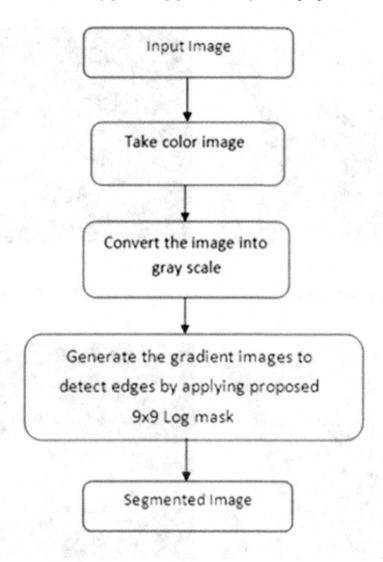

entropy are provided below (Table 1). Entrop, PSNR, and MSE measurements from gradient images generated using both the proposed 9x9 Log mask for Lena and Bird and the traditional 5x5 LoG mask are also displayed in the figures below from Figure 29 to Figure 31.

Figure 26. Original images: (a) Lena, (b) bird

Figure 27. Images of gradients produced with a typical 5x5 LoG mask: (a) Lena, (b) bird

CONCLUSION

One of the key processes in image processing, image analysis, image pattern recognition, and computer vision approaches is edge detection. However, significant (and productive) research has also been conducted recently on computer vision techniques that do not directly rely on edge detection as a pre-processing step. For

Figure 28. Images of gradients produced using the proposed 9x9 LoG mask: (a) Lena, (b) bird

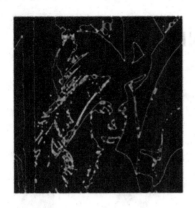 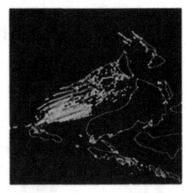

Table 1. Statistical analysis

IMAGE	METHOD USED	ENTROPY	PSNR	MSE
Fig. 27: (a) Lena.	with 5x5 LoG mask	5.0245	6.2983	1.5249e+004
Fig. 27: (b) Bird.	with 5x5 LoG mask	4.3404	5.7677	1.7231e+004
Fig. 28: (a) Lena.	with 9x9 LoG mask	5.2249	6.4520	1.4719e+004
Fig. 28: (b) Bird.	with 9x9 LoG mask	5.3574	5.8109	1.7060e+004

Figure 29. Entropy measurement from Gradient images obtained using standard 5x5 LoG maskand proposed 9x9 Log mask for Leena and bird

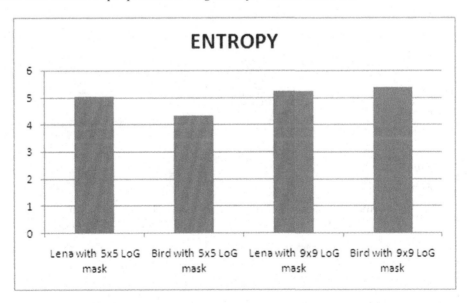

Figure 30. PSNR measurement from Gradient images obtained using standard 5x5 LoG maskand proposed 9x9 Log mask for Leena and bird .

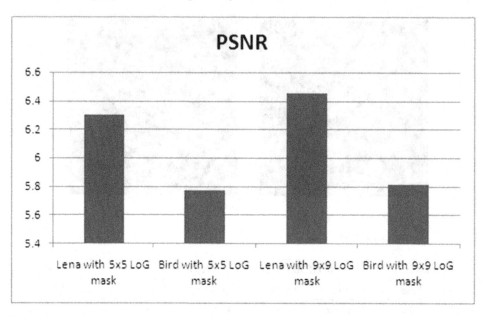

Figure 31. MSE measurement from Gradient images obtained using standard 5x5 LoG maskand proposed 9x9 Log mask for Leena and bird

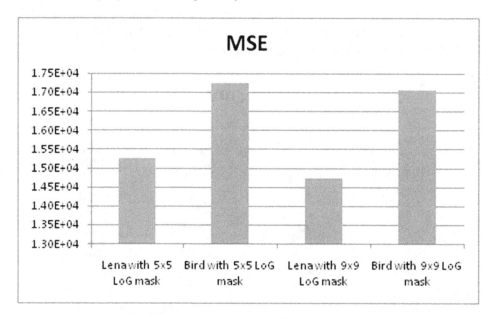

the analysis and measurement of several fundamental qualities associated with an object or objects of an image, such as area, perimeter, and form, correct identification of the edges of image objects is crucial. The segmentation method for natural photos developed in the present study effectively employs the idea of edge detection with gradients. When compared to the traditional 5x5 Laplacian and Gaussian filter for generating gradient images, the suggested approach provides resultant images with 9x9 Laplacian and Gaussian (LOG) filters that have a substantially higher accuracy to detect object edges. MATLAB software is used to compare the effectiveness of various edge detection approaches on two photos.

REFERENCES

Akkus, Z., Galimzianova, A., Hoogi, A., Rubin, D. L., & Erickson, B. J. (2017). Deep learning for brain mri segmentation: State of the art and future directions. *Journal of Digital Imaging*, *30*(4), 1–11. doi:10.100710278-017-9983-4 PMID:28577131

Alice Nithya, C. Lakshmi. (2015). Iris recognition techniques: A Literature Survey, Project: Multi-Unit Feature Level Fusion Approach Using PPCA, July 2015. *International Journal of Applied Engineering Research: IJAER*, *10*(12), 32525–32546.

Arinaldi, A., Pradana, J., & Gurusinga, A. (2018). Detection and classification of vehicles for traffic video analytics. *Procedia Computer Science*, *144*, 259–268. doi:10.1016/j.procs.2018.10.527

Aznag, K., Datsi, T., El oirrak, A., & El bachari, E. (2020). Ahmed El oirrak & Essaid El bachari (2020), Binary image description using frequent itemsets. *Journal of Big Data*, *7*(1), 32. doi:10.118640537-020-00307-8

Banerjee, S., & Mitra, S. (2020). Novel Volumetric Sub-region Segmentation in Brain Tumors. *Frontiers in Computational Neuroscience*, *14*, 3. doi:10.3389/fncom.2020.00003 PMID:32038216

Ben Aziza, S., Dzahini, D., & Gallin-Martel, L. (2015). A high speed high resolution readout with 14-bits area efficient SAR-ADC adapted for new generations of CMOS image sensors. *2015 11th Conference on Ph.D. Research in Microelectronics and Electronics (PRIME)*, (pp. 89-92). 10.1109/PRIME.2015.7251341

Chawla, S., & Oberoi, A. (2011). *Algorithm for Iris Segmentation and Normalization using Hough Transform*. International Conference on Advanced Computing and Communication TechnologiesAt: Panipat, Haryana, India.

Chen, H.-T., Wu, Y.-C., & Hsu, C.-C. (2016). Daytime Preceding Vehicle Brake Light Detection Using Monocular Vision. *IEEE Sensors Journal, 16*(1), 120–131. doi:10.1109/JSEN.2015.2477412

Cootes, T., Beeston, C., Edwards, G., & Taylor, C. (1999). Unified Framework for Atlas Matching Using Active Appearance Models. *Lecture Notes in Computer Science, 1613*, 322–333. doi:10.1007/3-540-48714-X_24

Davis, L. S., Rosenfeld, A., & Weszka, J. S. (1975). Region extraction by averaging and thresholding [J]. *IEEE Transactions on Systems, Man, and Cybernetics, 1975*(3), 383–388. doi:10.1109/TSMC.1975.5408419

Debelee, T., Schwenker, F., Rahimeto, S., & Ashenafi, D. Y. (2019). Evaluation of modified adaptive k-means segmentation algorithm. *Computational Visual Media, 5*(4), 347–361. doi:10.100741095-019-0151-2

Deepthi, S. (2021, June). Sandeep Kumar, Dr. Suresh L. (2021), Detection and Classification of Objects in Satellite Images using Custom CNN [IJERT]. *International Journal of Engineering Research & Technology (Ahmedabad), 10*(06).

El Shair, Z., & Rawashdeh, S. A. (2022). High-Temporal-Resolution Object Detection and Tracking Using Images and Events. *Journal of Imaging, 8*(8), 210. doi:10.3390/jimaging8080210 PMID:36005453

Ganesan, P., & Sajiv, G. (2017). A comprehensive study of edge detection for image processing applications. *International Conference on Innovations in Information, Embedded and Communication Systems (ICIIECS),* (pp. 1-6). IEEE. 10.1109/ICIIECS.2017.8275968

Santos, J., Pyrcz, M., & Prodanović, M. (2022). 3D Dataset of binary images: A collection of synthetically created digital rock images of complex media. *Data in Brief, 40*. doi:10.1016/j.dib.2022.107797

Jesorsky, O., Kirchberg, K. J., & Frischholz, R. W. (2001). Robust Face Detection Using the Hausdorff Distance. In J. Bigun & F. Smeraldi (Eds.), Lecture Notes in Computer Science: Vol. 2091. *Audio- and Video-Based Biometric Person Authentication. AVBPA 2001.* Springer. doi:10.1007/3-540-45344-X_14

Kamra, P., Vishraj, R., & Kanica, S. G. (2015). Performance Comparison of Image Segmentation Techniques for Lung Nodule Detection in CT Images. *International Conference on Signal Processing, Computing and Control (ISPCC),* (pp. 302-306). IEEE. 10.1109/ISPCC.2015.7375045

Kapoor, L., & Thakur, S. (2017). A Survey on Brain Tumor Detection Using Image Processing Techniques. *2017 7th International Conference on Cloud Computing, Data Science & Engineering – Confluence*. IEEE.

Kong, H., Akakin, H. C., & Sarma, S. E. (2012). A Generalized Laplacian of Gaussian Filter for Blob Detection and Its Applications. *IEEE Transactions on Cybernetics*, *43*(6), 1719–1733. doi:10.1109/TSMCB.2012.2228639 PMID:23757570

Kumar, G., Rampavan, M., & Paul Ijjina, E. (2021). Deep Learning based Brake Light Detection for Two Wheelers. *2021 12th International Conference on Computing Communication and Networking Technologies (ICCCNT)*, (pp. 1-4). IEEE. 10.1109/ICCCNT51525.2021.9579918

Kuo, C., Cheng, S., Lin, C., Hsiao, K., & Lee, S. (2017). *Texture-based Treatment Prediction by Automatic Liver Tumor Segmentation on Computed Tomography*. IEEE.

Lee, T.-H., Chou, H.-S., Chen, T.-Y., Lo, W.-S., Zhang, J.-T., Chen, C.-A., Lin, T.-L., & Chen, S.-L. (2022). Laplacian of Gaussian Based on Color Constancy Algorithm for Surrounding Image Stitching Application. *IEEE International Conference on Consumer Electronics*, (pp. 287-288). IEEE. 10.1109/ICCE-Taiwan55306.2022.9869055

Liu, Y., Shi1, C., Lin, B., Ha, C., Papanikolaou, N. (2009). Delivery of four-dimensional radiotherapy with Track Beam for moving target using an AccuKnife dual-layer MLC: Dynamic phantoms study. *Journal of Applied Clinical Medical Physics*, *10*(2), 2926. doi:10.1120/jacmp.v10i2.2926 PMID:19458594

Mahmoud, F. & Al-Ahmad, H. (2016). Two dimensional filters for enhancing the resolution of interpolated CT scan images. *2016 12th International Conference on Innovations in Information Technology (IIT)*, (pp. 1-6). IEEE. 10.1109/INNOVATIONS.2016.7880034

Moghbel, M., Mashohor, S., Mahmud, R., & Iqbal Bin Saripan, M. (2016). Automatic liver tumor segmentation on computed tomography for patient treatment planning and monitoring. *EXCLI Journal*, *15*, 406–423. PMID:27540353

Mugahed, A. (2018). A fully integrated computer-aided diagnosis system for digital X-ray mammograms via deep learning detection, segmentation, and classification. *International Journal of Medical Informatics*, *117*, 44–54. doi:10.1016/j.ijmedinf.2018.06.003 PMID:30032964

Murthy, T. S. D., & Sadashivappa, G. (2014). Brain tumor segmentation using thresholding, morphological operations and extraction of features of tumor. *2014 International Conference on Advances in Electronics Computers and Communications*. IEEE. 10.1109/ICAECC.2014.7002427

Muthukrishnan, R., & Radha, M. (2011). Edge Detection Techniques For Image Segmentation [IJCSIT]. *International Journal of Computer Science and Information Technologies*, *3*(6).

Nguyen, B. P., Heemskerk, H., So, P. T., & Tucker-Kellogg, L. (2016). Superpixel-based segmentation of muscle fibers in multi-channel microscopy. *BMC Systems Biology*, *10*(S5, Suppl 5), 124. doi:10.118612918-016-0372-2 PMID:28105947

Cootes, T.F., Lindner, C., Carmona, I.T., & Carreira, M.J. (2019). Fully Automatic Teeth Segmentation in Adult OPG Images. In: Vrtovec, T., Yao, J., Zheng, G., Pozo, J. (eds) Computational Methods and Clinical Applications in Musculoskeletal Imaging. Springer, Cham.] doi:10.1007/978-3-030-11166-3_2

Park, M., Jung, J., Oh, Y., & You, H. (2010). Assessment of Epicardial Fat Volume With Threshold-Based 3-Dimensional Segmentation in CT: Comparison With the 2-Dimensional Short Axis-Based Method. *Korean Circulation Journal*. .] doi:10.4070/kcj.2010.40.7.328

Pirhonen, J., Ojala, R., Kivekäs, K., Vepsäläinen, J., & Tammi, K. (2022). Brake Light Detection Algorithm for Predictive Braking. *Applied Sciences (Basel, Switzerland)*, *12*(6), 2804. doi:10.3390/app12062804

Priesnitz, J., Rathgeb, C., Buchmann, N., Busch, C., & Margraf, M. (2021). Buchmann (2021). An overview of touchless 2D fingerprint recognition. *EURASIP Journal on Image and Video Processing*, *8*(1), 8. doi:10.118613640-021-00548-4

Safia, D. (2016). *Batouche, M.* Quantum Genetic Computing and Cellular Automata for Solving Edge Detection.

Sathesh, A., Eisa, E., & Babikir, A. (2021), Hybrid Parallel Image Processing Algorithm for Binary Images with Image Thinning Technique. *Journal of Artificial Intelligence and Capsule Networks, 03*(3), 243-258. . doi:10.36548/jaicn.2021.3.007

Selvakumar, P., & Hariganesh, S. (2016). The performance analysis of edge detection algorithms for image processing. *International Conference on Computing Technologies and Intelligent Data Engineering (ICCTIDE'16)*, (pp. 1-5). IEEE. 10.1109/ICCTIDE.2016.7725371

Selvaraj, M. G., Vergara, A., Ruiz, H., Safari, N., Elayabalan, S., Ocimati, W., & Blomme, G. (2019). AI-powered banana diseases and pest detection. *Plant Methods, 15*(1), 92. doi:10.118613007-019-0475-z

Shah, A., Kulkarni, J., Patil, R., Sisode, V., & Gogave, S. (2020). Traffic Control System and Technologies: A Survey. *International Journal of Engineering and Technical Research, 9*(01). doi: . doi:0.17577/IJERTV9IS010246

Shanmathi, N., & Jagannath, M. (2018). Computerised Decision Support System for Remote Health Monitoring: A Systematic Review. *IRBM, 39*(5), 359-367.] doi:10.1016/j.irbm.2018.09.007

Sharma, N. (2014, June). Image Segmentation and Medical Diagnosis. *International Journal of Engineering Trends and Technology, 12*(2), 94–97. doi:10.14445/22315381/IJETT-V12P216

Singh, V., Sharma, N., & Singh, S. (2020). A review of imaging techniques for plant disease detection. *Artificial Intelligence in Agriculture*. https://doi.org/ 4, 229-242. doi:10.1016/j.aiia.2020.10.002.Volume

Yeom, S. (2011). *Augmented Reality for Learning Anatomy*. University of Tasmania.

Sun, Y., Ren, Z., & Zheng, W. (2022). Research on Face Recognition Algorithm Based on Image Processing. *Computational Intelligence and Neuroscience, 2022,* 9224203. Advance online publication. doi:10.1155/2022/9224203 PMID:35341202

Sun, Z. (2022). Application of Image Super-Resolution Reconstruction in Gymnastics Training by Using Internet of Things Technology. Computational Intelligence and Neuroscience. https://doi.org/ doi:10.1155/2022/8133187

Tahir, A., Munawar, H. S., Akram, J., Adil, M., Ali, S., Kouzani, A. Z., & Mahmud, M. (2022). M.A.P. Automatic Target Detection from Satellite Imagery Using Machine Learning. *Sensors (Basel), 22*(3), 1147. doi:10.339022031147 PMID:35161892

Tanzi, L., Piazzolla, P., Porpiglia, F., & Vezzetti, E. (2021). Real-time deep learning semantic segmentation during intra-operative surgery for 3D augmented reality assistance. *International Journal of Computer Assisted Radiology and Surgery, 16*(9), 1435–1445. doi:10.100711548-021-02432-y PMID:34165672

Teoh, K. H., Ismail, R. C., Naziri, S. Z. M., Hussin, R., Isa, M. N. M., & Basir, M. S. S. M. (2021). "Face Recognition and Identification using Deep Learning Approach" Journal of Physics: Conference Series PAPER, OPEN ACCESS, KH Teoh. *Journal of Physics: Conference Series, 1755*(1), 012006. doi:10.1088/1742-6596/1755/1/012006

Wu, W. (2016), Paralleled Laplacian of Gaussian (LoG) edge detection algorithm by using GPU. *Eighth International Conference on Digital Image Processing*. SPIE. 10.1117/12.2244599

Wu, Z. & Leahy, R. (1993). An optimal graph theoretic approach to data clustering: Theory and its application to image segmentation [J]. *IEEE transactions on pattern analysis and machine intelligence, 1993, 15*(11), 1101- 1113.

Yang, G., & Xu, F. (2011). Research and analysis of Image edge detection algorithm Based on the MATLAB. *Procedia Engineering, 15*, 1313–1318. doi:10.1016/j.proeng.2011.08.243

Zimeras, S. (2010). Segmentation Techniques of Anatomical Structures with Application in Radiotherapy Treatment Planning. *ACM Transactions on Graphics, 29*(5), 134. doi:10.1145/1857907.1857910

Zimeras, S. (2012). Segmentation Techniques of Anatomical Structures with Application in Radiotherapy Treatment Planning: Modern Practices in Radiation Therapy. *InTech*. doi:10.5772/34955

Chapter 5
Augmented Reality and Its Significance in Healthcare Systems

Ashish Tripathi
School of Computing Science and Engineering, Galgotias University, Greater Noida, India

Nikita Chauhan
G.L. Bajaj Institute of Technology and Management, Greater Noida, India

Arjun Choudhary
Sardar Patel University of Police, Security, and Criminal Justice, Jodhpur, India

Rajnesh Singh
School of Computing Science and Engineering, Galgotias University, Greater Noida, India

ABSTRACT

Augmented reality and virtual reality are terms often used together and even interchangeably sometimes without knowing their actual meaning. Augmented reality (AR) enhances the real world by mixing and overlapping digital objects with the real world whereas virtual reality (VR) is a completely different world created in a virtual space. VR can be experienced with wearables; but AR needs a device as simple as a phone and it's also wearable. In this chapter, AR is discussed and explored in a detailed manner. With its rapid evolving time, AR will be more common than it is now. It already is a part of everyone's life with the help of applications like Google Lens and Snapchat. AR has been experimented on for a while and the first spine surgery on a patient has been performed by John Hopkins neurosurgeon on June 8, 2020, using AR headsets. In this chapter, the types of extended reality (XR) and their differences. AR technology used in the study of anatomy, medical surgeries, pharma study, MedTech, and case studies of AR implementation in the field of medical surgeries is discussed.

DOI: 10.4018/978-1-6684-7659-8.ch005

INTRODUCTION

With the evolving industry, technology has improved a lot and things like augmented reality (AR) and virtual reality (VR) are not the technology of future anymore. Both the technologies have started integrating with many fields and industries including everyday life.

AR and VR terms are more commonly been used interchangeably by the people, but it all comes under Extended Reality (XR). With new types coming, the XR term was coined to represent any immersion in the real and virtual world. This technology still has a lot of potentials to be explored and might take a few more years to be completely a part of our life. It may be new but it has been around and worked upon since the mid-1950s. It was back in 1957, the filmmaker Morton Leonard Heilig, came up with a prototype to enhance the experience of watching films and theatres. He made a big box-like machine where he played five films and the viewer would experience different senses like visual, sound, smell, and touch. He named it Sensorama (Jones 2018,183). The viewer using the Sensorama would feel the motion, smell, touch, and realistic visuals along with the film to make the experience as real as possible. This Sensorama was not easily portable and needed space to use.

Later in 1968, a computer scientist, Ivan Edward Sutherland, created the very first head-mounted display to make the experience more portable. This display was similar to the new wearables but was not practical enough. It was a level up from Sensorama but the head mount display was heavy to be used, hence it was not a successful production but made a mark in the field.

In the mid-1970s, another computer artist, (Myron Krueger 1985), created a virtual interface called 'Videoplace'. This was a grade-up from previous inventions as it was more interactive. It used goggles and gloves to enhance the experience. It was not portable but rather needed a good amount of space to execute. It was different because it was not interconnecting a digital film with humans but two actual humans can interact without physically being in the same room. Videoplace needed two rooms for users to experience it. Each participant is present in two different rooms and while using a projector, cameras, gloves and goggles. A silhouette image of each other was visible to both through which they would interact. Along with this. they can also interact with virtual objects integrated by the programmer.

Around the 1980s is when portable and wearable devices started to stem. The heads-up displays were researched and experimented with by many researchers around the world.

One of the notable creations was of Steve Mann, an engineer who made the first Wearable computing' (Parisay 2020). He made the first wearable computing device called 'EyeTap'. This is similar to the head-up device and current google glass. In this, a wearable glass with cameras, index-matching and beam splitter is used to display the computer information to the user (Parisay 2020). It not only shows the digital information to the user but also records the user's perspective of the user in a computer.

Later in the 80s many people made different prototypes, integrating the digital world and real world. These inventions were taking place without any terms dedicated to it. Myron Krueger did name it "Artificial reality" but it was Jaron Lanier, a computer scientist, who popularly named it "virtual reality".

Later in 1990, Thomas P. Caudell, a researcher, coined the term "augmented reality" which made a brief difference between two different types of reality being developed.

TYPES OF EXTENDED REALITY (XR)

Extended reality itself is not referred to any particular type of reality but is just an umbrella term that includes any type of real and virtual world immersion. It is divided into three major types i.e., augmented reality (AR), virtual reality (VR), and mixed reality (MR). (Figure 1).

This division is based upon the reality-virtuality continuum. Paul Milgram (1995) was the first to propose and defined this reality-virtuality continuum in the early 90s. This reality-virtuality continuum is a spectrum showing the ranges of virtuality and the reality environment being immersed.

In Figure 1, the extreme left of the spectrum represents the real environment with no digital immersion at all. As we go to the right, the virtual immersion increases with the extreme right representing the completely virtual environment which represents the virtual reality. Augmented reality lies on the middle left of the continuum and on the middle right lies partial virtual reality or augmented virtuality.

Figure 1. Types of extended reality

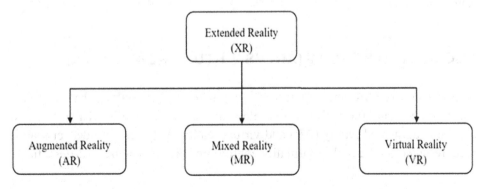

Augmented reality (AR) refers to the overlaying of digital objects into other real worlds which can be seen and accessed using devices. The real world dominates the environment and the virtual environment is in minority. Whereas in augmented virtuality (AV), the real-world objects are mapped into the virtual environment, partially or completely. Everything in between the two extreme ends of the reality-virtuality continuum is termed mixed reality (MR) (Milgram 1995).

Augmented reality (AR) and augmented virtuality (AV) become the subtypes of it. Mixed reality refers to the reality where both real and virtual objects partially contribute and are mixed to create a new and different environment. This reality-virtuality continuum is one of the most relevant and used concepts to differentiate different realities in existence (Speicher 2019).

MIXED REALITY (MR)

Mixed Reality in the reality-virtuality continuum lies between the middle of the ends of the spectrum. Paul Milgram and Fumio Kishino introduced the term "mixed reality" in "A Taxonomy of Mixed Reality Visual Displays" published in 1994 (Skarbez 2021) (Fig. 2). Except for the virtual environment and real world, mixed reality refers to and covers all other types of reality. Augmented reality has many evident examples including smart glasses to a very simple and common Snapchat app. An environment in which digital objects are spawned into the real world and are observed through a screen. Whereas augmented virtuality (AV) being on the right side of the continuum has more of the virtual environment where real-world objects or information are transported into.

The majority of the things are virtual in this type of reality. Weather forecasting channels are one the examples using augmented virtuality (AV). In this, the host's real image is overlaid on the digital image of the maps and the host continues to report (Olmedo 2013). This superimposition of real image over the synthetic image in which the virtual environment dominates comes under Augmented virtuality (Olmedo 2013).

AUGMENTED REALITY (AR) VS VIRTUAL REALITY (VR)

These two terms (AR and VR) generally get mixed and everything that integrates digital objects and the real world is referred to as virtual reality. The difference between augmented reality (AR) and virtual reality (VR) lies a lot deeper than it is represented on the reality-virtuality continuum (Fig 2). The basic differentiator of both realities is the dominant environment. In augmented reality, the real world dominates, and digital objects are overlaid and visible through a screen.

Whereas in virtual reality, there is no real world but only a completely different virtual world simulation created. These virtual worlds are entered and experienced through wearable devices. Table 1 shows the significant difference between augmented reality and virtuality reality.

AR basically enhances the real world by mapping digital objects. Something as simple as a phone can be used to experience augmented reality. There are many AR apps that help to give users a better experience. For example, Ikea has an app where one can spawn a piece of furniture in their room and see how it looks through the phone screen. This makes shopping easier and better.

There are many other such apps for glasses, clothes, makeup, etc. Augmented reality can be achieved and programmed using different methods i.e., marker-based

Figure 2. Reality-Virtuality continuum

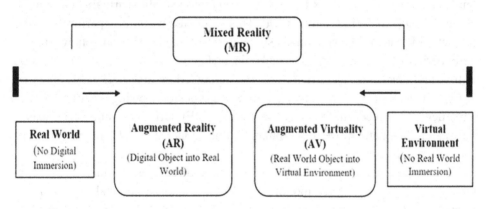

Table 1. Comparative analysis of augmented reality (AR) and virtuality reality (VR)

Augmented Reality	Virtual Reality
Enhances real world	Replaces real world
Combines digital objects with real world	Completely virtual world with all digital objects and no real-world information
Can see both real world and virtual object in one frame	Cannot see real world. Only virtual object
Has only 3 DOF (Degree of Freedom)	Has both 3 DOF and 6 DOF (Degree of Freedom)
Majority of real world and minority of virtual objects	No real-world object
Less expensive than VR	It is costly
Can be experienced using phone or smart glasses	Can be experienced using head-mounted display only
Ex – google glass, Pokémon Go	Ex- Oculus Rift, HoloLens

AR and marker-less AR. As the name suggests, in marked-based AR, there will be some sort of mark like QR code or target images, which is used to overlay digital objects in a given space.

Any image is overlaid on that particular mark; hence it should constantly be scanned by the camera. Marker-less AR does not require any marker to display digital information through the screen. It scans the room or given space dynamically by differentiating the floor and wall corners of the room. These are useful in an unpredictable environment where the space for digital information to be displayed is not specified and can be any size.

Marker-less AR programming uses SLAM (simulation localization and mapping). SLAM helps map suitable space by scanning the environment and creating a geometry to overlay digital objects. It uses devices' hardware and location to function. Marker-less-based AR has four further categories. Location-based AR, Projection-based AR, Superimposition AR, and Outlining AR. Each type is used for a specific type of virtual display (Frese 2010).

Virtual reality (VR) takes the experience into a completely different dimension than that of the real world. VR is the extreme end of the reality-virtuality continuum and has no objects from the real world. This simulated world is experienced using headsets. VR can also be generated using a multi-projected environment where one can experience and interact with virtual objects (Feng 2010). The level of interaction with virtual objects in VR is more than in augmented reality (AR).

One of the measurements that differentiate types of viewing in virtual reality is the degree of freedom. Degree of freedom (DOF) defines the number of axes a body can interact and move in a simulated dimensional space. There are two types of degrees of freedom i.e., 3 DOF and 6 DOF. 3 degrees of freedom refers to the access to three axes in space. X, Y, and Z are the three axes in which a subject in a simulated world can create motion. 3 DOF allows only rotational motion. After wearing a VR headset to enter the virtual world, they can only move their head left to right or up and down.

Only the head rotational motion is possible on the axes. Some of the VR headsets that support 3 DOF are google cardboard, Oculus Go, Google daydream, etc. 6 DOF allows more movement giving a better experience. It adds 3 more motions to the 3 DOF making a total of 6 accessible movements. 6 DOF permits translational motion along with the rotational motion. The person wearing a headset and additional haptic equipment can move around the simulated space in any direction. VR headsets that support 6 DOF are Oculus Rift, HTC Vive, Oculus quest 2, etc.

Figure 3. Head-Mounted Display

CURRENT TRENDS

The augmented reality (AR) industry has come far from its first origin. From the first big heavy box device used that was made to experience a different reality to portable AR devices. Today using AR became simple and we can experience it by just wearing AR glasses or even without it by just having a phone. Smart glasses are a growing industry and integrating everything into small parts of glass becomes a task. One of the most anticipated glasses is google glasses which are set to be launched in public soon.

Virtual reality (VR) devices are more complex than AR devices because they simulate another world. It is also portable enough but not as light as AR glasses. VR devices are more developed and have been doing well in the market. One of the leading virtual reality companies is Oculus. It makes headsets to play games and watch movies. Oculus quest is one of the most recent and popular headset versions. Oculus is also the first company to enter the VR market and develop headsets.

Others in the race are HTC, Microsoft, etc. also one of the popular headsets is HTC Vive and Microsoft HoloLens. Using these headsets one can watch movies and even play outdoor games like badminton inside without breaking anything. VR sets are easily accessible but these are expensive, hence not so commonly found everywhere.

Augmented reality (AR) is also used to create avatars of humans or other things. Holographic augmented reality helps to create real-life size digital objects. This has been used in many movies, videos, presentations, and even concerts. Virtual meetings are held using this technology with different avatars representing each member in a simulated room that can be entered using headsets.

AR technology is also majorly used in indoor and outdoor navigation systems. Google maps use AR to track live views while navigating. Integrating these technologies with navigation systems helps to navigate easily around large public areas like hospitals, campuses, airports, hospitals, etc. One can also backtrack the path it came from if they lose track.

AR has also made a mark in the automobile industry. One of the leading companies Tesla is using AR technology in their cars. With the help of artificial intelligence and augmented reality they are able to create more safe and accurate autopilot cars. Tesla has an AR view which shows the real-time view of what cars see while on the road. They have also integrated head-mounted displays which help to reduce any possible accident with the help of cameras around the car. All these technologies combined with AR reduce road accidents and also increase the safety of autopilot cars and other cars in general.

Other major companies like Toyota, and Mercedes-Benz, also started using augmented reality in their products. These head-mounted displays on cars can also be used to advance the parking lot system using 5G connectivity. Using 5G technology in cars will help to connect the car with the entering infrastructure and the available parking spots will be displayed on the head-mounted screen attached to the cars.

Metaverse is also another term that made people curious about AR and VR. In October 2021, Mark Zuckerberg rebranded Facebook to Meta and also announced that its upcoming work will be towards Metaverse. This is not the first time the word Metaverse has been coined.

The concept has been around since the 90s and the word was first introduced in a 90's fiction shows Snow Crash by Neal Stephenson. Despite being around for decades it was never so commonly discussed until the Facebook CEO publicly announced to invest in it. The Metaverse is a virtual simulated world in which one can enter and do activities.

According to Mark Zuckerberg, in the Metaverse many people can collectively enter and will be represented by their avatars. They can buy goods, clothes or even make virtual conference rooms. It will be a whole new world but virtual. The exact idea is still vague but the Metaverse will be a big step towards the future of extended reality.

AUGMENTED REALITY IN MEDICAL: SURGERY AND ANATOMY STUDY

Augmented reality has great potential in the medical field. It can make studying medicine much better and in-depth using AR. The advanced technology of AR can be used to perform complex surgeries more accurately. Augmented reality can render different images and this can be used to study anatomy in more detail. Having a 3D projected model of anatomy can be easily studied and analyzed. Not only for education but any patient's anatomy can be analyzed using similar ways. Using haptics, 3D visualization of anatomy can be controlled. Studying human anatomy is one of the most important parts of practicing medicine and performing future surgeries.

Traditionally it is done on cadavers to practice dissection and surgeries. It is still a better way but for multiple sessions of practice, AR is proven to be a better and more efficient alternative to it. As for using virtual reality to study anatomy, it will give a whole 360 visualization of the anatomy and can be studied in a completely different environment while being in a room with headsets. They can control the model and have more interactive study. The setup can be expensive but the quality of education will effectively increase. This will help to provide a better pathological understanding of the human body to find and perform surgery.

One of the applications that have been created to study anatomy is the magic mirror (Blum 2012). The magic mirror is an AR system where cameras are used with a motion sensor device Kinect, developed by Microsoft. To make this work, a person with a camera and Kinect attached to him will stand in front of the TV screen (Khor 2016).

The mirror image of the person will be displayed on the screen. The screen will show the CT scan of the person's body along with the body which creates the illusion of looking inside the body. It can be scrolled and the parts can be manipulated and controlled to better understand the structure. The CT data can be manipulated using a gesture-based user interface (Kamphuis 2014).

Augmented reality can also help to visualize dynamic organs like lungs. Hamza-Lup et al. in their paper mentioned a system that they developed that showed the real-time visualization of the lungs of a patient (Hamza-Lup 2007). What makes it different is that it was superimposed on a patient in an operating room (Kamphuis 2014) This was viewed using a head-mounted display. For the visualization, the patient's data was extracted using high-resolution computed tomography (HRCT). This data was combined with a standard dynamic lung model which created a patient-specific model (Kamphuis 2014).

This was overused to analyze the lungs of a patient which shows dynamic behaviour. This technology can also be used to train and educate the students. Dynamic behaviour of lungs like breathing patterns can be diagnosed more accurately.

Augmented reality combined with this help in surgeries like lung transplant and lung volume. It becomes easier and more accurate to analyze a patient's condition with the visualizing techniques provided by AR and VR.

Practicing surgeries can be done easily using AR. the overall efficiency of the successful surgery will increase. It has shown great application use in the Dental and Neurosurgery field of medicine. Neurosurgery is a complex surgery and needs to be done with more practice and accuracy.

AR can help to get hands-on practice multiple times before working on an actual patient. Neurosurgery involves a very small and specific part of the brain that needs to be treated and worked upon.

Augmented reality models can help to identify the specified area and the surgery can be performed without having to spend time ascertaining the part. This also helps in not accidentally damaging or hampering other parts of the same vicinity. Augmented reality has been used to advance the visualization of the liver for better understanding the structure for liver surgeries (Soler 2014).

Augmented reality has been used for several years to train medical students on surgeries such as blood clot removal or penis implant surgery. However, moving from training to regular use in surgery is taking a bit longer to be adopted. The University of Alabama with Emory University piloted orthopedic shoulder replacement using Google Glass, and Stanford University is developing its own device.

AR in Pharmaceuticals Industry

Augmented reality has been used for visualizing drugs in detail. AR or VR can be used to model a 3D version of the drug molecular structure while developing or discovering any drug (Ventola 2019). It will help to study it in more detail. During drug identification and making, understanding the molecular structure and the way it binds with other molecules within the drug becomes complex to comprehend considering its size. Visualizing tools using AR and VR will help to understand how macromolecule ligands bind with complex structures like proteins (Ratamero 2018, Liu 2018).

As AR is proven to be able to visualize dynamic human parts like lungs, it is also able to visualize other dynamic structures. Protein is of dynamic nature which tends to change its structure with a little change in its configuration. If even one amino acid is changed then the overall structure and behaviour of the protein will change. Despite this dynamic behaviour and structure of the protein, Augmented reality or virtual reality can help to visualize it in real-time. Different mutations of protein in the drug discovery can be displayed in an understandable way. This not only helps to visually observe but also can be virtually simulated to see changes and mutations (Ratamero 2018, Liu 2018). These changes can be reversed as it is done digitally in a virtual environment. This helps to do as much trial and error as we want before actually working on the drug design.

Virtual reality technology can also provide the solution to any issues that may occur and is found in the visualized model due to its nature. VR can help to find efficient ligands for drug composition better than manual methods (Ventola 2019).

As much as it is useful in drug discovery and research, AR and VR technology is helpful in making pharmacist education and training more interactive and better.

As a student in pharmacy, understanding the composition and interaction of drugs is a complex thing to learn. Regular class sessions may find it hard to explain the 3D and dynamic nature of molecules. Visualization tools of AR make it easy to understand the drug-receptor interaction more easily (Krueger 1985). It promotes more active learning among students.

They practice different techniques like compounding medication (Ventola 2019) and other procedures. These technologies are very advanced, and efficient and would increase the overall quality of education but it is costly and is not as common in university programs.

CASE STUDIES

The chapter discusses two major applications and case studies of augmented reality in healthcare. There are many experiments done using augmented reality and virtual reality. Many of these few only made a mark in the medical field and are effectively used by others. One of them is spine surgery performed by 8 neurosurgeons from John Hopkin Hospital.

An AR guidance system was made by a MedTech manufacturer Augmedics in collaboration with John Hopkin Hospital. Augmedics designed and manufactured an AR headset and the hospital neurosurgeons used it to perform a very crucial spine surgery successfully. It was first done on a cadaver and then on an actual patient. Another case study is of a MedTech start-up AccuVein. It is an AR device that is designed to visually map veins on the person's skin surface. It turns out to be a successful MedTech start-up and has been used a lot. These two had made a significant mark in using augmented reality in the medical field efficiently.

John Hopkins Surgery

John Hopkin Hospital used the potential of augmented reality in two of its surgeries. One of them was a spinal surgery that took place on June 8, 2020. They treated the excruciating pain in the lower back and leg caused by arthritis. To perform the surgery three neurosurgeons worked together. They inserted six screws and rods in the spine of the patient to fuse three vertebrae which helped to relieve the pain. This spinal decompression procedure is done by surgeons without any headsets but just with their own eyes and X-ray on the screen for position reference. To advance the surgery and make it easier and more accurate, John Hopkins neurosurgeons used AR headsets to perform it. These AR headsets displayed the visual images of the patient's internal anatomy based on the patient's X-ray and CT scans. These images showed bones and tissues right in front of the surgeon's eyes which helped them perform the surgery easily without having to constantly look at another screen with X-rays of the patient.

As the technology used is augmented reality, they were also able to see the patient in the real world as the images were see-through from the headset and overlaid it in the real world. These headsets also worked as guides to place the screws by showing the path and position where the screws need to be placed. This surgery was successful only because the headsets were made so efficiently. "When using augmented reality in the operating room, it's like having a GPS navigator in front of your eyes," said Dr. Timothy Witham, director of the Johns Hopkins Neurosurgery Spinal Fusion Laboratory. The gear was manufactured by Augmedics, a medical technology manufacturer based in Illinois, US. This see-through AR headset was named "xvision Spine System", which became the first augmented reality guidance system for surgery.

The John Hopkins neurosurgeons tested the surgery with this AR headset on a cadaver. After successful placement of screws on the spine, they performed the surgery on a woman and further on many other patients. This surgery marked the first successful spine surgery using augmented reality. Other than this, John Hopkins neurosurgeons also performed another surgery on June 10, 2020. This was also a spine surgery using the same "xvision Spine System". For this surgery, a cancerous tumor, chordoma was removed from the patient's spine. It was also a successful surgery marking another benchmark in treating tumors using AR. The "xvision Spine System" is used in many more spine surgeries after that.

AccuVein (start-up)

AccuVein is a MedTech start-up that made a medical imaging solution. This is a device that helps to visibly map veins on the skin's surface. It uses projection-based AR to map the veins. It is a portable device that began to be used in many health departments like dermatology, emergency department (ED), etc. It is based in New York, US. Using this over the skin will show veins as black lines over a red infrared light square box on the skin surface. The device uses two low-power lasers. One is 642 nm red laser light and another is 785 nm infrared laser (Aulagnier 2014). The reason for the veins being displayed in black is because hemoglobin absorbs infrared light, showing veins as black lines without much detailed display of the vein's walls. One of the reasons why AccuVein is a success in the market is because of its portability and non-invasive techniques which makes it safe to use even on children.

One of the case studies shows the effective use of AccuVein in pediatric dermatology where skin lesions were diagnosed. An atrophic plaque was observed on a 5-months old boy. Later when diagnosed with this device, a large vein with a small superficial branching vessel network was found near the atrophic plaque (AlZahrani 2019). This new identification of the vein turned out to be very useful as the child was then diagnosed with RICH, a no involuting congenital hemangioma. It was treated successfully and discharged after a year. AccuVein helped to diagnose the vein without any invasion as it is important in pediatric dermatology to have less invasion but more accuracy. This case used an infrared AccuVein AV400 vein finder. This projection-based AR device was made with a laser-based scanner, processing system, infrared red light, etc. which makes the visibility of the underlying vasculator possible by creating a virtual real-time image. AccuVein was also featured in a Harvard Business review article "A Manager's Guide to Augmented Reality" by Michael E. Porter, reviewing the use of AR in healthcare.

CONCLUSION

Augmented reality and virtual reality are advanced technologies that are still emerging into different fields. Back from the 50's when the first experiment on creating a different reality, we've come a long way. That vague concept got classified into distinct types. Only after the late 80s and early 90s, Jaron Lanier and Thomas P. Caudell coined the terms "virtual reality" and "Augmented reality" which made other researchers and engineers to work and grow it individually.

Augmented reality (AR) enhances the real world by emerging digital objects into it. Whereas Virtual Reality replaces the real world and is a complete digital simulation creating a virtual world. AR has become more common and is used in apps like Snapchat and google Lenses. Augmented reality can be experienced using something as simple as phones or using smart glasses which are also being developed by big tech companies.

Virtual reality (VR), on the other hand, needs a head-mounted display (HMD) to experience the virtual world created. Paul Milgram and Fumio Kishino introduced the reality-virtuality continuum and Mixed Reality (Fig. 2) in "A Taxonomy of Mixed Reality Visual Displays" published in 1994. This RV-continuum became the most cited in future papers and also became a standard spectrum to classify different types of realities. AR and VR grew and different industries started using them in their fields and products. Automobiles, entertainment, education, medical, etc. have incorporated this technology to advance the industry. Companies like Tesla have successfully used AR in their cars to advance the auto-pilot function and reduce road accidents. Games can be played at home using HMDs like Oculus Rift, HTC Vive, etc. The 6 DOF (Degree of Freedom) in VR increased the experience along with haptics for hand movements. In the healthcare field, Augmented Reality has been used to study anatomy better, drug discovery and understand the small molecular structure, or even practice surgeries virtually instead of a cadaver.

AR has also found its way in helping to treat cancerous tumors like chordoma and performing spine surgery. Augmedics, a MedTech start-up along with John Hopkin Hospital successfully performed spine surgery using augmented reality in a headset named "xvision Spine System". Another MedTech start-up AccuVein used AR to make a device that is used to map veins on the skin surface. It is a small portable device that has been successfully used. Many experiments took place to emerge this high-end technology in the medical field but these two made a mark and are a huge success. Augmented reality and virtual reality are two technologies that are still not used to their full potential but are still growing.

AR and VR get interchangeably used but they are very different in many terms. After the recent announcement of Facebook to rebrand its name to "meta" and focusing its upcoming projects on metaverse, people started to show interest in AR and VR. The metaverse started being discussed and people became curious to know more about it and the technology to be used. In no time it will be as common as phones are today.

REFERENCES

AlZahrani, F., Crosby, M., & Fiorillo, L. (2019, March 1). Use of AccuVein AV400 for identification of probable RICH. *JAAD Case Reports, 5*(3), 213–215. doi:10.1016/j.jdcr.2018.11.022 PMID:30809562

Aulagnier, J., Hoc, C., Mathieu, E., Dreyfus, J. F., Fischler, M., & Le Guen, M. (2014, August). Efficacy of AccuVein to facilitate peripheral intravenous placement in adults presenting to an emergency department: A randomized clinical trial. *Academic Emergency Medicine, 21*(8), 858–863. doi:10.1111/acem.12437 PMID:25176152

Blum T, Kleeberger V, Bichlmeier C, & Navab N. (2012). Mirracle: An augmented reality magic mirror system for anatomy education. In *2012 IEEE Virtual Reality Workshops* (pp. 115-116). IEEE.

Feng, Y., Duives, D. C., & Hoogendoorn, S. P. (2022, March 1). Development and evaluation of a VR research tool to study wayfinding behaviour in a multi-story building. *Safety Science, 147*, 105573. doi:10.1016/j.ssci.2021.105573

Frese, U., Wagner, R., & Röfer, T. (2010, September). A SLAM Overview from a User's Perspective. *Kunstliche Intelligenz, 24*(3), 191–198. doi:10.100713218-010-0040-4

Hamza-Lup, F. G., Santhanam, A. P., Imielinska, C., Meeks, S. L., & Rolland, J. P. (2007, January 2). Distributed augmented reality with 3-D lung dynamics—A planning tool concept. *IEEE Transactions on Information Technology in Biomedicine*, *11*(1), 40–46. doi:10.1109/TITB.2006.880552 PMID:17249402

Jones, S., & Dawkins, S. (2018). The sensorama revisited: evaluating the application of multi-sensory input on the sense of presence in 360-degree immersive film in virtual reality. In *Augmented reality and virtual reality* (pp. 183–197). Springer. doi:10.1007/978-3-319-64027-3_13

Kamphuis, C., Barsom, E., Schijven, M., & Christoph, N. (2014, September). Augmented reality in medical education? *Perspectives on Medical Education*, *3*(4), 300–311. doi:10.1007/S40037-013-0107-7 PMID:24464832

Khor, W. S., Baker, B., Amin, K., Chan, A., Patel, K., & Wong, J. (2016, December). Augmented and virtual reality in surgery—the digital surgical environment: Applications, limitations and legal pitfalls. *Annals of Translational Medicine*, *4*(23), 454. doi:10.21037/atm.2016.12.23 PMID:28090510

Krueger, M. W., Gionfriddo, T., & Hinrichsen, K. (1985, April). VIDEOPLACE—an artificial reality. In *Proceedings of the SIGCHI conference on Human factors in computing systems* (pp. 35-40).

Liu, X. H., Wang, T., Lin, J. P., & Wu, M. B. (2018, December 2). Using virtual reality for drug discovery: A promising new outlet for novel leads. *Expert Opinion on Drug Discovery*, *13*(12), 1103–1114. doi:10.1080/17460441.2018.1546286 PMID:30457399

Milgram, P., Takemura, H., Utsumi, A., & Kishino, F. (1995). Augmented reality: A class of displays on the reality-virtuality continuum. In *Telemanipulator and telepresence technologies*, *2351*, 282-292. Spie.

Olmedo, H. (2013, January 1). Virtuality continuum's state of the art. *Procedia Computer Science*, *25*, 261–270. doi:10.1016/j.procs.2013.11.032

Parisay, M., Poullis, C., & Kersten, M. (2020). Eyetap: A novel technique using voice inputs to address the midas touch problem for gaze-based interactions. arXiv preprint arXiv:2002.08455. Feb 19.

Ratamero, E. M., Bellini, D., Dowson, C. G., & Römer, R. A. (2018, June). Touching proteins with virtual bare hands. *Journal of Computer-Aided Molecular Design*, *32*(6), 703–709. doi:10.100710822-018-0123-0 PMID:29882064

Richardson, A., Bracegirdle, L., McLachlan, S. I., & Chapman, S. R. (2013, February 12). Use of a three-dimensional virtual environment to teach drug-receptor interactions. *American Journal of Pharmaceutical Education, 77*(1), 11. doi:10.5688/ajpe77111 PMID:23459131

Skarbez, R., Smith, M., & Whitton, M. C. (2021, March 24). Revisiting milgram and kishino's reality-virtuality continuum. *Frontiers in Virtual Reality, 2*, 647997. doi:10.3389/frvir.2021.647997

Soler, L., Nicolau, S., Pessaux, P., Mutter, D., & Marescaux, J. (2014, April). Real-time 3D image reconstruction guidance in liver resection surgery. *Hepatobiliary Surgery and Nutrition, 3*(2), 73. PMID:24812598

Speicher, M., Hall, B. D., & Nebeling, M. (2019). What is mixed reality? In Proceedings of the CHI conference on human factors in computing systems. ACM. doi:10.1145/3290605.3300767

Ventola, C. L. (2019, May). Virtual reality in pharmacy: Opportunities for clinical, research, and educational applications. *P&T, 44*(5), 267. PMID:31080335

Chapter 6

Automated Diagnosis of Eye Problems Using Deep Learning Techniques on Retinal Fundus Images

N. Sasikaladevi
SASTRA University (Deemed), India

S. Pradeepa
SASTRA University (Deemed), India

K. Malvika
SASTRA University (Deemed), India

ABSTRACT

Automated diagnosis of eye diseases using deep learning techniques on retinal fundus images has become an active area of research in recent years. The suggested method divides retinal images into various disease categories by extracting relevant data using convolutional neural network (CNN) architecture. The dataset used in this study consists of retinal images taken from patients with various eye conditions, such as age-related macular degeneration, glaucoma, and diabetic retinopathy. The aim of this study is to investigate the potential of deep learning algorithms in detecting and classifying various retinal diseases from fundus images. The suggested approach may make early eye disease diagnosis and treatment easier, reducing the risk of vision loss and enhancing patient quality of life. The DenseNet-201 model is tested and achieved an accuracy rate of 80.06%, and the findings are extremely encouraging.

DOI: 10.4018/978-1-6684-7659-8.ch006

1. INTRODUCTION

The general workflow of automated diagnosis of eye diseases using deep learning techniques on retinal fundus images involves several steps. Firstly, retinal fundus images are preprocessed to remove noise, enhance contrast, and standardize their size. Then, CNN-based algorithms are trained on large datasets of labelled retinal fundus images to learn the patterns and features of different eye diseases. During the training process, the CNN algorithm learns to classify retinal fundus images into different disease categories. The trained model is then validated on a separate set of images to evaluate its performance. Metrics such as sensitivity, specificity, and area under the receiver operating characteristic curve (AUC-ROC) are used to measure the accuracy of the model. The model can be used to automatically diagnose eye problems once it has been trained and validated. Retinal fundus images are input to the trained model, and the output of the algorithm most likely provides the diagnosis of the disease. In some circumstances, the algorithm can also provide a likelihood score for each type of sickness, indicating how confident the diagnosis is. Diabetic eye disease (DED) is a group of eye problems that can affect diabetic people. Over time, diabetes can harm the eyes, resulting in blurry vision or even total blindness. Therefore, it is crucial to identify DED signs early in order to stop the condition from progressing and receive prompt treatment. (Mule et al., 2019), (WebQiao et al., 2020) (Li et al., 2020).

2. RELATED WORK

The KNN model reported by (Singh et al., 2022). achieved an accuracy of 99%, which is not the best result. (Sesikala et al., 2022) CNN .'s model and (Akbar et al., 2022) almix's of DarkNet and DenseNet both showed increased accuracy. The highest levels of accuracy have been delivered by KNN, DarkNet, and DenseNet. Nevertheless, (Suganyadevi et al., 2022) CNN's model could not be considered the most accurate since it only managed an average accuracy of 85%. For the other models, the accuracy ranges were 89.29% to 98%.

3. PROPOSED ARCHITECTURE

The effectiveness and precision of ophthalmic diagnosis could be greatly increased by automating the diagnosis of eye illnesses using deep learning methods on retinal fundus images. The following is a suggested structure for such a system: Data gathering: Gathering a sizable dataset of retinal fundus photographs is the initial

Table 1. Literature

Authors	Model	Accuracy
(Singh et al., 2022)	KNN	99%
(Sarki et al., 2020)	AlexNet	97.93%
(Shoukat et al., 2021)	GoogleNet	97.8%
(Akbar et al., 2022)	DarkNet + DenseNet	99.7%
(Sesikala et al., 2022)	CNN	99.89%
(Qureshi et al., 2021)	ADL-CNN	98%
(Suganyadevi et al., 2022)	CNN	85%
(Kaushik et al., 2021)	Stacked Convolutional Neural Network	97.92%
(Gupta et al., 2022)	Inception V3	92%

stage. This can be accomplished utilizing either image collections from current databases or brand-new image acquisition employing dedicated fundus cameras. Pre-processing: The pictures must be preprocessed after the dataset is acquired to remove any potential artefacts or noise. Some preprocessing techniques include Image cropping, resizing, color normalizing, and contrast boosting. Data Augmentation: Techniques like rotation, flipping, and noise addition can be used to expand the dataset and enhance the deep learning model's capacity for generalization. Model Training: The pre-processed and enhanced dataset can be used to train a deep learning model, such as a convolutional neural network (CNN). To classify retinal fundus images into several disease categories, such as diabetic retinopathy, glaucoma, and age-related macular degeneration, the model needs to be trained. Model evaluation: A different test dataset should be used to gauge how well the trained model performed. The performance of the model can be assessed using evaluation measures including accuracy, precision, recall, and F1 score. Clinical Validation: After the model has been trained and assessed, it can be put to the test in a clinical environment to gauge how well it performs in actual situations. This may entail contrasting the model's automated diagnosis with the diagnosis offered by ophthalmologists. Deployment: The model may be used in clinical settings to help ophthalmologists make diagnoses of eye problems if it performs well in clinical validation. The proposed framework, in its entirety, entails gathering and pre-processing a sizable dataset of retinal fundus images, training a deep learning model to classify the images into various disease categories, assessing the model's performance, clinically validating the model, and deploying the model in a clinical setting.

3.1. DenseNet-201

As an addition to the DenseNet-121 and DenseNet-169 architectures, Huang et al. presented the DenseNet-201 deep convolutional neural network architecture in 2017. The primary concept of DenseNet is to feed-forward connect each layer to every other layer, which helps to solve the vanishing gradient issue and encourages feature reuse. Fig. 1 shows the DenseNet Architecture.

DenseNet connects each layer to all succeeding layers in a dense block, resulting in a dense connection pattern, as opposed to standard convolutional neural networks, in which each layer is simply connected to the preceding layer. The architecture is made up of several dense blocks that are joined by transition layers that condense the feature maps' spatial dimensions. Many convolutional layers, a batch normalisation layer, and a Rectified Linear Unit (ReLU) activation function make up each dense block in DenseNet-201. A dense connectivity pattern is then produced by concatenating the output of each dense block with the output of every preceding block. The channel axis is used during the concatenation operation, allowing the network to reuse features that were learned by prior layers. A batch normalisation layer, a 1x1 convolutional layer, and a 2x2 average pooling layer make up DenseNet-201's transition layers. The average pooling layer and 1x1 convolutional layer are utilised to minimise the feature maps' spatial dimensions and channel count, respectively. The last layers of DenseNet-201 are a fully connected layer that generates the final classification output and a global average pooling layer that calculates the average of each feature map

Figure 1. DenseNet architecture

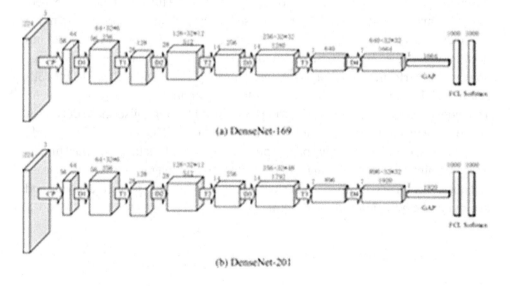

(a) DenseNet-169

(b) DenseNet-201

across its spatial dimensions. The network can have a fixed-size output for any input image size thanks to the global average pooling layer, which makes it more suited for deployment on various devices. Average pooling and maximum pooling are the two types of pooling techniques utilised in DenseNet201.A feature map is compressed by a factor of two using average pooling, which takes the average of a set of values. On the other hand, max pooling decreases the size of a feature map by a factor of two by taking the largest value of a group of data. DenseNet-201 is a powerful deep learning architecture for image processing tasks. Using several image classification benchmarks, including the ImageNet dataset, a sizable image classification dataset with over 1 million images and 1000 classes, it has been demonstrated to produce state-of-the-art results. Compared to conventional convolutional neural networks, DenseNet-201's dense connectivity design enables it to extract more complicated characteristics from the input photos. This is so that greater feature reuse and increased network efficiency are made possible by each layer having access to the features that were learned by all the preceding layers. To increase performance on a particular task, it has also been applied in transfer learning, where a pre-trained model is adjusted on a fresh dataset. Object identification, semantic segmentation, and picture classification are just a few of the tasks that DenseNet-201 may be used for. To enhance performance on a particular image classification job, a pre-trained DenseNet-201 model, for instance, can be fine-tuned on a smaller dataset. Overall, DenseNet-201 has produced state-of-the-art results on numerous benchmarks and is a potent tool for deep learning-based image processing. It is an excellent candidate for a variety of image processing jobs because of its dense connection structure, which allows it to extract more complicated characteristics from input images.

3.2. Transfer Learning

Transfer learning and fine-tuning are commonly used techniques in machine learning for improving the performance of models.

These techniques can also be applied in the context of medical diagnosis. Using a pre-trained model that has been trained on a sizable dataset for a new but similar job is called transfer learning. For instance, a model for classifying medical photos that have already been trained on a sizable dataset of natural images can be utilized as a starting point for the training process. Fig. 2 shows the transfer learning model. Fine-tuning is a technique that involves taking a pre-trained model and continuing its training on a new dataset. In the context of medical diagnosis, fine-tuning could involve taking a pre-trained model that has been trained on natural images and continuing its training on a dataset of medical images. . Fig. 3 shows the transfer learning model from source model to target model.

Figure 2. Transfer Learning from source model to target model

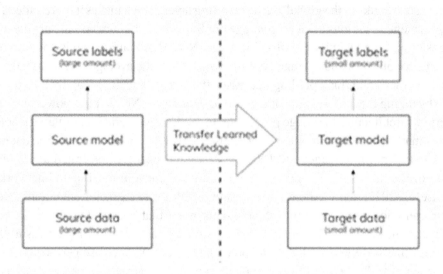

Figure 3. Layer description in transfer learning

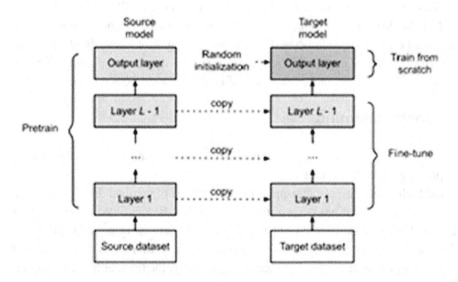

The pre-trained model is utilised as a starting point for the fine-tuning process, and the model's weights are updated by training on the new dataset. This technique enables the model to maintain the broad information it gained from the first dataset while adapting to and learning the specific features of the new dataset. When the target dataset is small and it is impractical to train a model from scratch, fine-tuning is especially helpful. Because the previously trained model serves as a useful starting point for learning the new task, fine-tuning also aids in lowering the computing costs associated with model training. Combining these methods can be especially helpful in the field of medical diagnostics, since datasets are frequently small and getting new data can be expensive and time-consuming. A model can be trained on a large dataset of natural images and then fine-tuned on a smaller dataset of medical images using transfer learning and fine-tuning, which results in better performance than developing a model from scratch on the medical images alone. In image diagnosis, a layer graph can be used to represent the structure of a neural network model used for image classification or segmentation. A layer graph is a visual representation of the layers in a neural network and the connections between them. Fig. 4 depicts the layer graph for the proposed model.

Figure 4. Layer graph

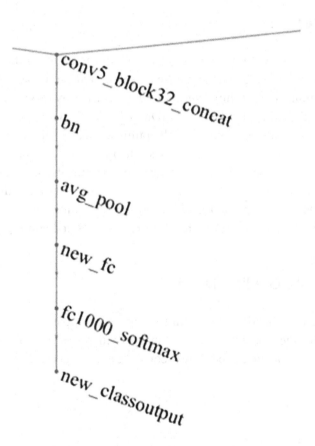

In a typical image diagnosis task, the input image is first fed into a convolutional neural network (CNN) that is designed to extract relevant features from the image. The CNN typically consists of several layers of convolutional and pooling operations, followed by one or more fully connected layers.

4. EXPERIMENTAL SETUP

4.1 System Configuration

The computer system is equipped with an Intel i5 11th gen processor with integrated graphics, 3.20 GHz, 6 Core(s), 12 Logical Processor(s), 16GB (2*8) DDR4 RAM, a 512GB M.2 PCIe solid-state drive. The operating system installed is Windows 11 Home 64-bit. It also features a Gigabit Ethernet port, Wi-Fi 6 connectivity, Bluetooth v5.1, and a range of USB 3.2 Gen 1 and Gen 2 ports. MATLAB Version: 9.13.0.2105380 (R2022b) Update.

4.2 Optimizers

The basic idea behind optimization is to compute the gradients of the loss function concerning the model parameters and use these gradients to update the parameter values. The optimizer algorithm determines the specific way in which the gradients are used to update the parameters. SGDM (Stochastic Gradient Descent with Momentum), Adam, and RMSprop are all optimization algorithms commonly used in deep learning for image processing tasks. The most effective option depends on the particular problem being solved and the features of the data being used because each of these optimisation techniques has its own advantages and disadvantages. Prior to the selection of the optimal optimisation algorithm for a given task, it is frequently a good idea to test out a few alternative ones and compare their results on a validation set.

4.2.1 Stochastic Gradient Descent

This is a basic optimization algorithm that updates the parameters of the neural network based on the gradient of the loss function with respect to those parameters. It is simple and efficient, but can be slow to converge and can get stuck in local minima.

Figure 5. Stochastic gradient descent validation

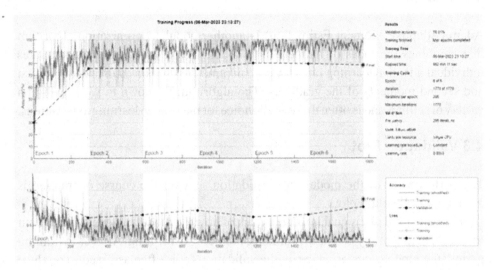

4.2.2 Root Mean Square Propagation

RMSprop (Root Mean Square Propagation) is a variant of gradient descent that also uses adaptive learning rates, but instead of estimating the second moment of the gradients like Adam, it computes a moving average of the squared gradients. This helps smooth out the optimization trajectory and prevent oscillations in the gradient updates.

Figure 6. RMSprop validation

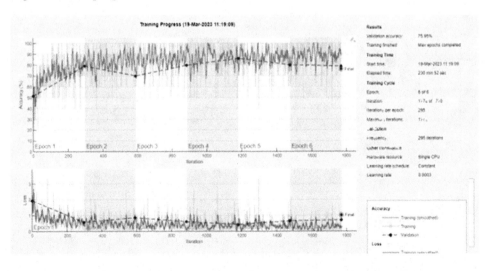

4.2.3 Adam

Adam (Adaptive Moment Estimation) is another popular optimization algorithm that combines the ideas of momentum and adaptive learning rates. It computes individual adaptive learning rates for each parameter based on estimates of the first and second moments of the gradients. The algorithm is known to work well in a variety of settings and is often the default choice for many deep learning applications.

4.3 Validation Plot

The performance of the model on a validation set over the course of training is plotted in a validation plot.

The validation plot's goal is to keep track of how the model performs throughout training and to spot any potential problems, including overfitting. After each training period, the model's performance on the validation set is often assessed to produce a validation plot. Afterwards, a plot of the validation accuracy or loss against the quantity of training epochs is shown. The resulting plot can be used to identify possible problems like overfitting and to see how the model's performance evolves over time. As a result, it's crucial to routinely produce validation plots throughout the training process in order to assess the model's effectiveness and make any necessary modifications.

Figure 7. Adam optimizer validation

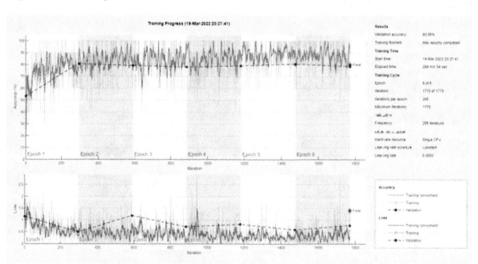

Figure 8. Sample validation data set

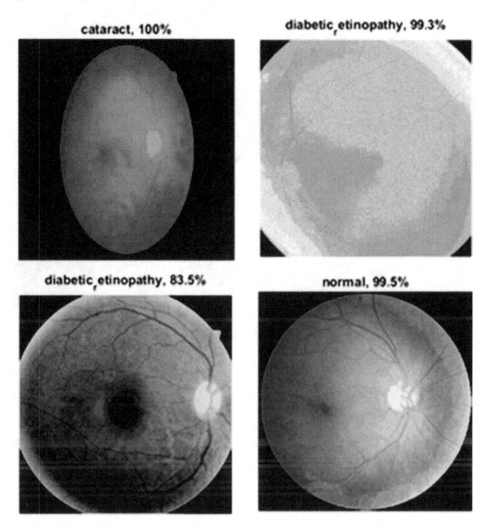

4.4 Confusion Chart

In deep learning for image processing, a confusion chart (or confusion matrix) is a table that is used to evaluate the performance of a classification model. The table shows the number of correct and incorrect predictions made by the model on a set of test data, broken down by class. The confusion chart offers a thorough analysis of the model's performance for each class, making it a useful tool for assessing a classification model's effectiveness. Potential problems including unbalanced datasets, instances where one class may be overrepresented in the training data, or

Figure 9. Confusion matrix

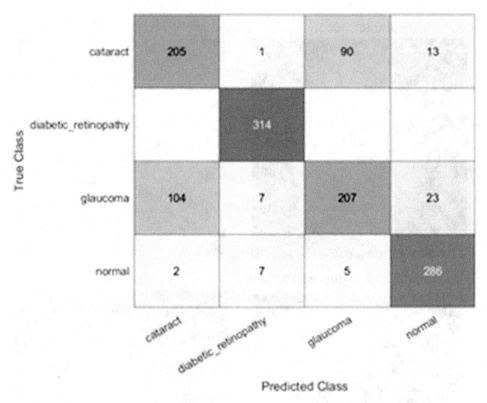

instances where the model is predisposed to favour one class over the others can all be found using this method.

4.5 AUC-ROC Curve

The AUC-ROC curve is a common evaluation metric used in deep learning for image classification tasks. AUC stands for Area Under the Curve, while ROC stands for Receiver Operating Characteristic. The AUC-ROC curve is a plot of the True Positive Rate (TPR) against the False Positive Rate (FPR) for different classification thresholds. The TPR represents the proportion of true positive instances that are correctly identified by the model, while the FPR represents the proportion of false positive instances that are incorrectly identified by the model. The ideal curve for a deep learning model is one that hugs the top-left corner of the plot and has a high TPR and low FPR. With a higher AUC value signifying greater performance, the AUC-ROC curve is a single number that summarises the model's overall performance across all conceivable classification criteria. The TPR and FPR trade-off for various classification thresholds is visualised by the AUC-ROC curve, making it a valuable tool for measuring the performance of a deep learning network. It is especially helpful when the positive class is uncommon since it can be used to determine the appropriate threshold for increasing the TPR while lowering the FPR.

5. CONCLUSION

In conclusion, the use of the Densenet201 model and the Adam optimizer has shown promising results in the analysis of retinal fundus images. Accurate classification of

Figure 10. AUC-ROC curve for the model

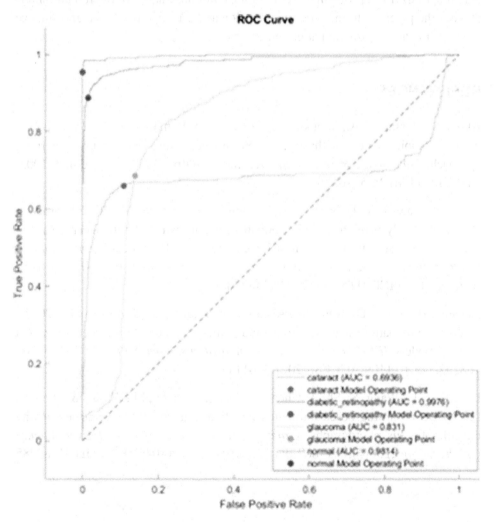

retinal fundus images can be accomplished by utilising a deep learning strategy with a potent convolutional neural network architecture like Densenet201 and an effective optimizer like Adam. It has been demonstrated that the Densenet201 model can efficiently collect characteristics at different levels of abstraction in retinal fundus images, enabling it to spot patterns suggestive of different eye diseases. The use of the Adam optimizer has also been shown to improve the training efficiency of the model, leading to faster convergence and improved accuracy. Overall, the use of retinal fundus images to diagnose eye problems has the potential to be greatly enhanced by the combination of Densenet201 and Adam optimizer. However, further research is needed to investigate the performance of the model on larger and more diverse datasets and to explore the generalizability of the model across different clinical settings and populations. In conclusion, the Densenet201 model with the Adam optimizer is a promising method for examining images of the retinal fundus and has the potential to increase the precision and effectiveness of diagnosing eye illnesses further improving patient outcomes.

REFERENCES

Akbar, S., Hassan, S. A., Shoukat, A., Alyami, J., & Bahaj, S. A. (2022). Detection of microscopic glaucoma through fundus images using deep transfer learning approach. *Microscopy Research and Technique*, *85*(6), 2259–2276. doi:10.1002/jemt.24083 PMID:35170136

Gupta, S., Panwar, A., Kapruwan, A., Chaube, N., & Chauhan, M. (2022, February). Real Time Analysis of Diabetic Retinopathy Lesions by Employing Deep Learning and Machine Learning Algorithms using Color Fundus Data. In *2022 International Conference on Innovative Trends in Information Technology (ICITIIT)* (pp. 1-5). IEEE. 10.1109/ICITIIT54346.2022.9744228

Kaushik, H., Singh, D., Kaur, M., Alshazly, H., Zaguia, A., & Hamam, H. (2021). Diabetic retinopathy diagnosis from fundus images using stacked generalization of deep models. *IEEE Access : Practical Innovations, Open Solutions*, *9*, 108276–108292. doi:10.1109/ACCESS.2021.3101142

Li, F., Yan, L., Wang, Y., Shi, J., Chen, H., Zhang, X., Jiang, M., Wu, Z., & Zhou, K. (2020). Deep learning-based automated detection of glaucomatous optic neuropathy on color fundus photographs. *Graefe's Archive for Clinical and Experimental Ophthalmology*, *258*(4), 851–867. doi:10.100700417-020-04609-8 PMID:31989285

Mule, D. B., Chowhan, S. S., & Somwanshi, D. R. (2019). Detection and Classification of Non-proliferative Diabetic Retinopathy Using Retinal Images. In K. Santosh & R. Hegadi (Eds.), *Recent Trends in Image Processing and Pattern Recognition. RTIP2R 2018. Communications in Computer and Information Science* (Vol. 1036). Springer.

Qiao, L., Zhu, Y., & Zhou, H.WebQiao. (2020). Diabetic Retinopathy Detection Using Prognosis of Microaneurysm and Early Diagnosis System for Non-Proliferative Diabetic Retinopathy Based on Deep Learning Algorithms. *IEEE Access : Practical Innovations, Open Solutions*, 8, 104292–104302. doi:10.1109/ ACCESS.2020.2993937

Qureshi, I., Ma, J., & Abbas, Q. (2021). Diabetic retinopathy detection and stage classification in eye fundus images using active deep learning. *Multimedia Tools and Applications*, 80(8), 11691–11721. doi:10.100711042-020-10238-4

Sarki, R., Ahmed, K., Wang, H., & Zhang, Y. (2020). Automatic detection of diabetic eye disease through deep learning using fundus images: A survey. *IEEE Access : Practical Innovations, Open Solutions*, 8, 151133–151149. doi:10.1109/ ACCESS.2020.3015258

Sesikala, B., Harikiran, J., & Sai Chandana, B. (2022, April). A Study on Diabetic Retinopathy Detection, Segmentation and Classification using Deep and Machine Learning Techniques. In *2022 6th International Conference on Trends in Electronics and Informatics (ICOEI)* (pp. 1419-1424). IEEE.

Shoukat, A., Akbar, S., Hassan, S. A. E., Rehman, A., & Ayesha, N. (2021). An automated deep learning approach to diagnose glaucoma using retinal fundus images. In 2021 international conference on frontiers of information technology (FIT). IEEE.

Singh, L. K., Khanna, M., & Thawkar, S. (2022). A novel hybrid robust architecture for automatic screening of glaucoma using fundus photos, built on feature selection and machine learning-nature driven computing. *Expert Systems: International Journal of Knowledge Engineering and Neural Networks*, 39(10), e13069. doi:10.1111/ exsy.13069

Suganyadevi, S., Renukadevi, K., Balasamy, K., & Jeevitha, P. (2022, February). Diabetic Retinopathy Detection Using Deep Learning Methods. In *2022 First International Conference on Electrical, Electronics, Information and Communication Technologies (ICEEICT)* (pp. 1-6). IEEE. 10.1109/ICEEICT53079.2022.9768544

Chapter 7

Comparative Analysis and Automated Eight–Level Skin Cancer Staging Diagnosis in Dermoscopic Images Using Deep Learning

Auxilia Osvin Nancy V.

iD https://orcid.org/0000-0002-4254-0537

Department of Computer science and Engineering, College of Engineering and Technology, SRM Institute of Science and Technology, Vadapalani Campus, Chennai, India

P. Prabhavathy

Department of Computer science and Engineering, College of Engineering and Technology, SRM Institute of

Science and Technology, Vadapalani Campus, Chennai, India

Meenakshi S. Arya

Department of Transportation, Iowa State University, USA

B. Shamreen Ahamed

Deparment of Computer science and Engineering, College of Engineering and Technology, SRM Institute of Science and Technology, Vadapalani Campus, Chennai, India

ABSTRACT

The challenge in the predictions of skin lesions is due to the noise and contrast. The manual dermoscopy imaging procedure results in the wrong prediction. A deep learning model assists in detection and classification. The structure in the proposed handles CNN architecture with the stack of separate layers that use a differential function to transform an input volume into an output volume. For image recognition and classification, CNN is specifically powerful. The model was trained using labeled data with the appropriate class. CNN studies the relationship between input features and class labels. For model building, use Keras for front-end development and

DOI: 10.4018/978-1-6684-7659-8.ch007

Tensor Flow for back-end development. The first step is to pre-process the ISIC2019 dataset, splitting it into 80% training data and 20% test data. After the training and test splits are complete, the dataset has been given to the CNN model for evaluation, and the accuracy on each lesion class was calculated using performance metrics. The comparative analysis has been done on pretrained models like VGG19, VGG16, and MobileNet.

INTRODUCTION

Melanoma is the type of skin cancer that begins to grow out of control of the development of melanocytes (PS Staff, 2016). In this regard, the main factors for detecting skin cancer and distinguishing between benign and melanoma, such as symmetry, colour, size and shape (PS Staff, 2016). Many countries worldwide, especially the United States, report growing death rates from skin cancer (Marks, 1995). Recent cancer statistics and figures show that the calculable range of recent cancer cases of this kind is around 1.9 million and that in the United States the death rate will be around 608,570 (Siegel et al., 2023). Earlier diagnosis is likely to reduce the death rate. Daylight exposure is associated with the greatest risk of carcinoma development with every malignant melanoma and non-melanoma cancer in the skin. Current carcinoma encompasses malignant melanoma and NMSC malignancies, made up of basal (BCC) and squamous cell carcinoma (SCC), as indicated by Figure1 (Gordon, 2013). Melanoma is the deadliest type of cancer occurring in human beings that leads to coloured markings or skin moles. Clinical testing, dermoscopic image analysis, histological investigation, and ultimate biopsy are the initial diagnosis of carcinoma (Mane & Shinde, 2018).

Skin lesion classes:

Lesions images of skin are classified into seven classes.

Classification of Lesion images of Skin:

Figure 1. Skin cancer types

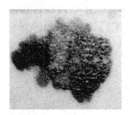 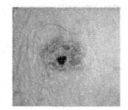 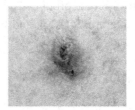

(a) Melanoma   (b) Squamous Cell Carcinoma   (c) Basal Cell Carcinoma

1. Basal Cell Carcinoma (BCC)
2. Benign Keratosis (BKL)
3. Actinic Keratos (AKIEC)
4. Dermato Fibroma (DF)
5. Melanoma (MEL)
6. Nevus (NV)
7. Vascular Lesion (VASC).

RELATED WORKS

Deep learning architectures are now used in medical image analysis and new frameworks are being developed to predict, diagnose and detect. Current neural networks that are good for image identification and outperform the categorization of skin cancer (Adegun & Viriri, 2020). The first job of classifying the image is to accept the image and how it is characterised by its class. The ability to recognise human images is quite different from the machine. The machine detects the images as pixels. CNN's are neural networks specially built for the detection of images and computer vision challenges. Unlike Machine Learning techniques, CNN uses a sequence of convolutionary, pooling and non-linear layer to process and pre-process the picture as a 2D vector and ultimately FCN (fully connected layer) to produce the output (Yamashita et al., 2018; Aghdam et al., 2017).

CNN-based profound learning algorithms have demonstrated amazing detection, classification and segmentation performance in medical imaging applications (Harley, 2015).

Zhang (2021) suggested a strategy for retrieving deep skin injury characteristics using deep CNNs. Models such as AlexNet, ResNet-18 and VGG 16 have been pre-trained. In the last stage, the features generated are provided to the SVM classification for training. Classifier used for classification purposes. The model was assessed using the 2017 ISIC dataset and the accuracy achieved was 97.55%.

Dorj et al. (2018) introduced the method of categorization by pretrained deep CNN of various types of skin lesion images (AlexNet used for feature extraction). The system provided produced the highest mean sensitivity, specificities, SCC precision, actinic keratosis (AK), and BCC values: 95.1% (98.9%), and 94.17% respectively.

METHODOLOGY

The framework deals with four pretrained model and one model with fine- tuned parameters of CNN. All the models gone through the training process. The pretrained models with weights are initialized from the dataset ImageNet and this has been done before the training on the dataset. The layers in the pretrained model are convolution layer, Batch normalization layer, ReLu, Maxpool, Fully connected, Dense and Softmax. In the proposed CNN fine-tuned model the sequential framework of the layers mentioned in pretrained model are used based on the requirements. Based on the validation data and frequency call all the models are trained using Adam optimizer at learning rate 0.001 with 30 epochs. The classification accordingly is splitted into two essential steps, pre-processing and classification. Initially, the publicly accessible ISIC2019 dataset is pre-processed by resizing the image. The pre-processed image is further to the fine- tuned CNN and pretrained model to do classification of skin images. Figure 2 shows the system flow of the proposed work (Jeyakumar et al., 2022).

SKIN LESION CLASSIFICATION-ANALYSIS ON DEEP LEARNING MODELS

Datasets

The computerized systems offered are utilized for diagnostics. The reliable collection of dermoscopic pictures is necessary to evaluate the model and determine diagnostic

Figure 2. System flow of skin cancer classification

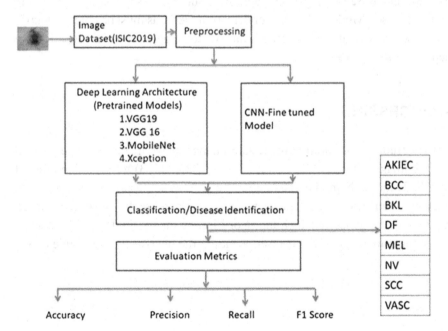

performance. It is essential that the dependable dataset is available. This section shows the data set available for the evaluation of the proposed detection approaches for skin cancer.

ISIC Archive

There are numerous data sets for cutaneous lesions in the ISIC collection (ISIC Archive, n.d). The International Skin Imaging Collaboration released the ISIC data set during the 2016 International Symposium on Biomedical Imaging (ISBI), also known as ISIC2016 (Razzak et al., 2020). Training and testing are two distinct areas of the ISIC2016 repository. ISIC contains 379 dermoscopic images in the test subset and 900 shots in the training subset.

The benign nevi, benign SK, and melanoma photo categories for the ISIC2017 collection. In the dataset, 2000 photographs are used for training, 150 photographs are used for validation, and 600 photographs are used for testing. 1372 photos of benign nevi, 1254 photos of SK, and 374 photos of melanoma make up the training data set. There are 30 pictures of melanoma and 78 pictures of benign nevi in this validation dataset. 42 SK pictures are included as well. The test data set contains 393 images of benign nevi, 90 photographs of SK, and 117 photographs of melanoma.

12,594 photos, 100 images, and 1000 images total are included in ISIC2018.

The eight categories of lesion images represented in the ISIC2019 dataset include melanoma, melanocytic-nevus, BCC, AK, BKIEC, DF, vascular lesion, and SCC. The test data set has 8239 images, along with an outlier class that wasn't present in the training data set. Modern skin cancer detection methods must be able to identify these images. Additionally, photo metadata like sex, age, and patient location are included in the ISIC2019 dataset.

PRE-PROCESSING

Resizing the Image: For fine-tuning the dataset, the images are resized. The original shape of the image is (450, 600, 3), it resized to (224,224, 3) for pretrained model and (28, 28, 3) for CNN model.

Image augmentation: The primary technique to increase the size of the dataset. The new images are generated, by rotating or flipping the original one, which helps to increase the dimension of the dataset, avoiding the overfitting of a single class.

Figure 3. Image augmentation

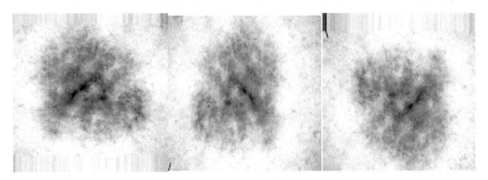

Pretrained Models

Model developed by another party to address a comparable issue is defined as pretrained. We start with the model that has been trained on another problem rather than creating a new model from scratch to overcome comparable problem. Training time will significantly improve with the addition of more convolutional blocks. Utilizing pre-trained models, so we would train a few layers rather than the full architecture.

The approach to fine-tuning the model is as follows.

Extraction of Features

The top layer should be get rid of (the one which gives the probabilities for being in 1000 classes).For the new data set, use the entire framework as a permanent feature extractor.

Utilize the Pre-Trained Model's Architecture

Utilize the model's architecture while randomly initializing each weight and retraining the model using the dataset.

Train Some Layers While Freezing Others

Keep the model's weight of first layer to be frozen and retrain the upper layers while training some layers and freezing others. Try and test the numbers of layers that should freeze.

Pretrained Model for Skin Lesion Classification

- VGG-19
- VGG-16
- MobileNet
- Inceptionv3 (GoogLeNet)
- ResNet50
- EfficientNet

VGG-19

CNN with 19 layers is called the VGG. Additionally, the ImageNet database's has the million photos higher are used to train network.1000 items can be classified using the network which is pretrained. The pixels with 224x224 colored pictures were used to train the network. It is distinguished by its simplicity by only taking into account three 3x3 convolutional layers stacked on top of one another with increasing depth. Reducing volume size is handled by max pooling. Two FC (completely connected) layers with 4,096 nodes each continue the layer soft max. In network the number "19" indicates weight of the layers. The architecture of VGG19 is shown in Fig.4 (Zheng et al., 2018).

The VGG-19 classification of skin lesions began with the database (ISIC 2019). The system flow of VGG 19 is shown in Figure 5 (Jaworek-Korjakowska et al., 2019). Initially, the dermoscopic images are preprocessed by resizing, frame removal, and skin lesion cropping. The second stage is classification, where preprocessed images are fed into the VGG-19 model for classifying the skin lesion. With the training dataset, the model was trained and validated by evaluation metrics.

The model is evaluated by the accuracy performed in testing and validation datasets. The classification report indicates the performance metrics as precision,

Figure 4. VGG19 architecture

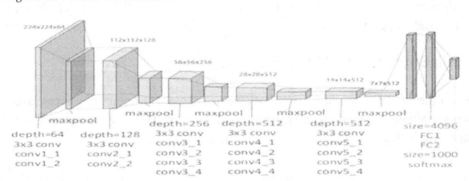

Figure 5. System flow of VGG19 model

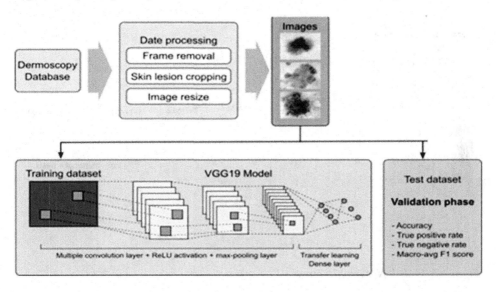

recall, and F1score. The skin lesion class prediction results in seven done by categorical entropy. The percentage of each class was calculated and visualised as predicted and expected skin lesion.

VGG-16

Convolution-followed-by-Maxpool layers are regularly placed across the whole architecture of the VGG-16 version of CNN. At the very end, it has two FC and Softmax outputs that are flawlessly connected. The number 16 in VGG16 stands for the weighted layers. With almost 138 million parameters, this is a huge network. The architecture of VGG-16 is shown in Figure 6 (Ren et al., 2018)

MobileNet Model

MobileNet are built with the depth wise separable convolutional layer. Each layer comprises in depth wise and a pointwise convolution. The network includes with 28 layers to make up MobileNet. By appropriately adjusting the width multiplier hyper parameter, the 4.2 million parameters in a conventional MobileNet can be further decreased. The supplied image is 224 by 224 by 3, in size. The architecture with specified layers is visualized in Figure 7 (Pujara, 2020).

Figure 6. Architecture of VGG16

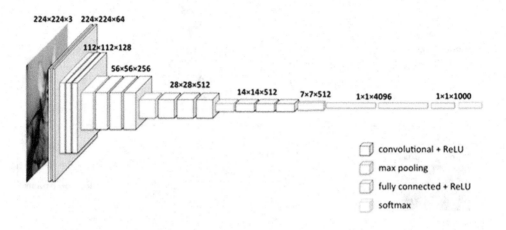

Figure 7. MobileNet architecture

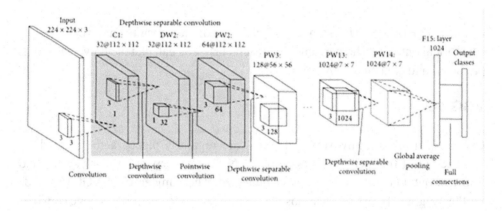

System Flow

The system flow of MobileNet is shown in Figure 8 (Chaturvedi et al., 2020).To perform classification, the skin lesion image from the dataset (ISIC, 2019) is disposed to the MobileNet model after removing the duplicates in the earlier process called preprocessing. Though the model is trained, it has the ability to classify the input image accordingly. The evaluation of the frame work was done by the validation dataset.

Figure 8. System flow for MobileNet model

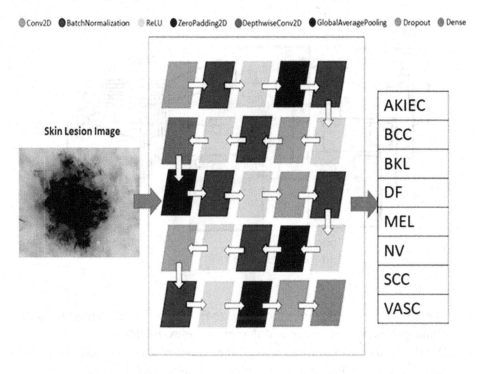

Xception

The filtering process simultaneously takes into account the Xception (extreme inception), the dimension (spatial), and the cross-channel or depth dimension. This is comparable to a convolutional filter scanning a 2x2 patch of pixels across all RGB channels at the input layer of an image. Using 1X1 convolution, the original input was projected onto multiple unique, smaller input spaces. We transformed those tiny 3D data blocks from each of those input environments using a different kind of filter. By taking it a step farther, Xception Instead of splitting the input data into numerous compressed chunks, it maps the spatial correlations for each output channel separately before executing an 11-depth convolution to capture cross-channel correlation. The architecture of the Xception model is visualized in Figure 9 (Xu, 2017).

CNN (Convolutional Neural Network) Model

The two basic components of CNN architecture.

Figure 9. Xception model

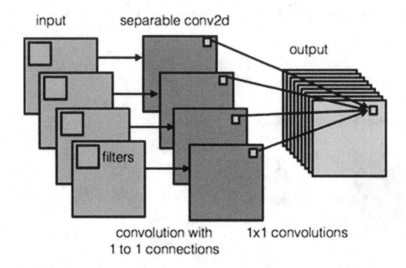

- A convolution tool that uses the "Feature Extraction" method to separate and catalogue the various aspects of the image for examination.
- A fully connected layer that utilises the output of the convolutional process and classifies the image using the previously retrieved features.

Convolutional, pooling, and fully connected layers make up CNN's three primary layers. The two more crucial characteristics, the drop out layer and activation function, are present in addition to this layer. Figure 10 (Prabhu, 2019) shows the CNN architecture.

Figure 10. CNN model

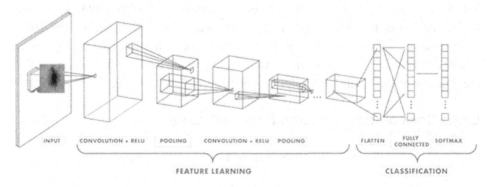

The proposed model is sequential with 7 convolution layers, 1 flattened and 4 dense layer (. The first layer is specified with input shape (28X28X3), kernel size, and activation function ReLu, followed by Max Pool. The two-dimensional array is flattened by the flattened layer to convert it into a single-dimensional array. The layer output finally enters the dense layer, specifying the number of output classes and activation function.The model for classifying the skin lesion follows the regular CNN (Convolution->Maxpool->Convolution->Maxpool) and it continues to perform feature extraction and classification.

MATHEMATICAL MODEL OF CNN

Convolution

Convolution alters our image based on the values of a small matrix of numbers (referred to as a kernel or filter) that is run over it. The formula below is used to generate the values of subsequent feature maps, where the input is represented by i and kernel by k. The result matrix's row and column indices are designated with a and c, respectively. The calculation is shown in equation 1.

$$G[a,c] = (n1 * k1)[a,b] = \sum_l \sum_m k1[l,m] i[a-l,c-k1] \tag{1}$$

Multiple Filters

If multiple filters used for same image, separate convolutions are performed, and the combined results are stacked one on top of the other.

The shape is calculated by

$$[i,i,n_{RGB}] * [f,f,n_{RGB}] = \left[\left[\frac{i+2i_p - f}{i_s} + 1 \right], \left[\frac{i+2i_p - f}{i_s} + 1 \right], n_f \right] \tag{2}$$

$i - imagesize, f - filtersize, n_{RGB} - number of channels, i_p - Padding, i_s - Stride$

Figure 11. Convolutional layer

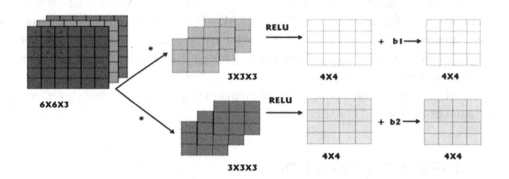

Example of One Convolutional Layer

The input image is 6X6X3 in size and multiplied to 3X3X3 with filters. Each of these filters turns on a different part of the images. Separate convolutions are performed, and the combined results are stacked one on top of the other. This is often called "activation" because only the features that have been turned on are passed on to the next layer. The convolution is determined by the input size and the filter. The output (convolution) size is 4X4 since it has an input of 6X6 and the filter is 3X3. The working of single convolution layer is visualized in Figure 11.

The number of parameters in one layer calculated by filters of a neural network. If the number of filters is 10 and the filter size is 3X3X3 then the number of parameters is 280. Figure 12 shows the visual representation for the parameter calculation.

Figure 12. Calculation of number of parameters in one layer

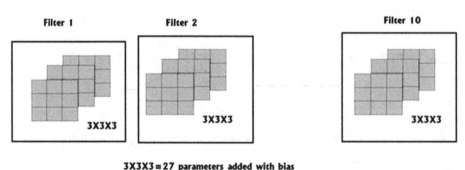

3X3X3 = 27 parameters added with bias
The total parameters for 1 filter = 28

The parameters for 10 filter = 280

Pooling Layer

The pooling layer performs nonlinear down sampling on the output to reduce the number of parameters that the network must learn. Each layer learns to distinguish various things as a result of the numerous repetitions of these stages.

TYPES OF POOLING

Max Pooling

The resultant matrix value of Max pool is calculated by the formula

$$Maxpool = \frac{i + 2i_p - f}{i_s} + 1 \tag{3}$$

Where, $i - inputsize,\ i_p - Padding, i_s - Stride, f - filtersize$

Average Pooling

The same procedure as Max pooling to identify the output matrix value. In this pooling consider the average value of the selected matrix.

RESULTS AND DISCUSSION

The pretrained architectures VGG-19, VGG-16, MobileNet, Xception, and fine-tuned CNN, which are utilised as the final classification layers in the fully connected layer, receive the dataset ISIC2019 as input data. This layer, which is the decision-making layer, also receives the categorization goal set. The only parameter that varies depending on the network utilised is the number of layers, which is the same for all models but varied for input size. The proposed parameters and performance measure are listed in Table 1 and 2.

Accuracy and Loss of Deep Learning Model

The training, validation accuracy and loss are visualized for all five architectures (ISIC Archive, n.d; Jaworek-Korjakowska et al., 2019; Razzak et al., 2020; Ren et al., 2018; Zheng et al., 2018).

Table 1. Five architectures: Proposed parameters

Parameters	Size of Input	Epoch	Weights	Optimizer	Classifier	Learning Rate
VGG-19	(224,224,3)	30	ImageNet	Adam	SoftMax	0.0001
VGG-16	(224,224,3)	30	ImageNet	Adam	SoftMax	0.0001
Xception	(224,224,3)	35	ImageNet	Adam	SoftMax	0.0001
MobileNet	(224,224,3)	15	ImageNet	Adam	SoftMax	0.0001
Fine-tuned CNN model	(28,28,3)	30	ISIC2019	Adam	SoftMax	0.0001

Table 2. Performance analysis of ISIC2019 dataset for the five architectures

ARCHITECTURE	PERFORMANCE MEASURES			
	TRAINING ACCURACY	VALIDATION ACCURACY	TESTING ACCURACY	F1 SCORE
VGG-19	0.6862	0.8448	0.8531	0.8201
VGG-16	0.6866	0.8392	0.8404	0.8331
Xception	0.83	0.8616	0.8397	0.83
MobileNet	0.94	0.8771	0.90	0.8761
Fine-tuned CNN model	0.96	**0.9207**	**0.9180**	**0.90**

Figure 13. VGG-19 model

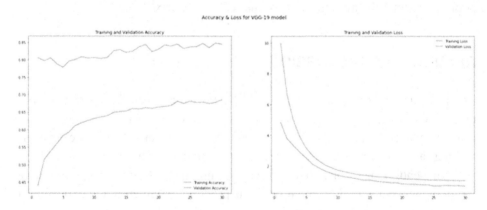

CONCLUSION AND FUTURE WORK

Deep learning techniques are efficient in identifying skin cancer earlier and are less expensive. Among five architectures, the fine-tuned CNN model obtained with a 92.07% work rate in the ISIC 2019 dataset is the best. In the future, real-time

Figure 14. VGG-16 model

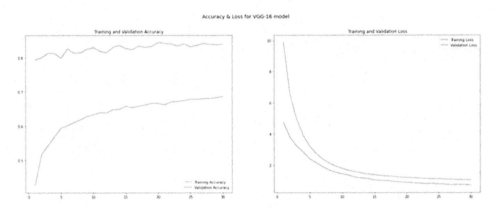

Figure 15. MobileNet model

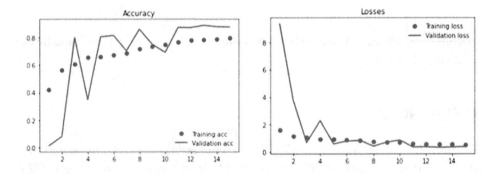

Figure 16. Xception model

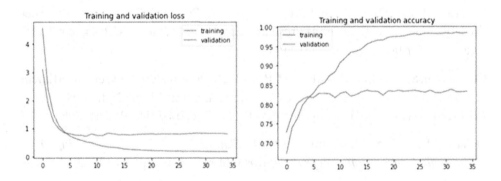

databases from various regions of skin and other attributes can be collected for clinical validation using a developed model. The methods can also be fine-tuned

Figure 17. Fine-tuned CNN model

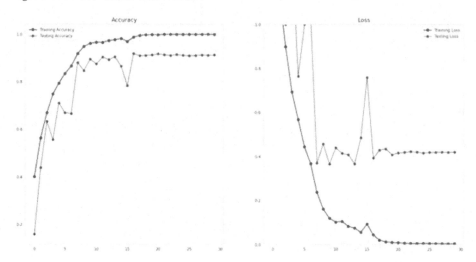

for clinicians. Though all methods are working well, further research should be the practical validation.

REFERENCES

Adegun, A. A., & Viriri, S. (2020). FCN-based DenseNet framework for automated detection and classification of skin lesions in dermoscopy images. *IEEE Access: Practical Innovations, Open Solutions*, 8, 150377–150396. doi:10.1109/ACCESS.2020.3016651

American Cancer Society. (n.d.). *Cancer facts & figures 2021*. American Cancer Society. https://www.cancer.org/research/cancer-facts-statistics/all-cancer-facts-figures/cancer-facts-figures-2021.html

Chaturvedi, S. S., Gupta, K., & Prasad, P. S. (2020). Skin lesion analyser: An efficient seven-way multi-class skin cancer classification using MobileNet. *Advances in Intelligent Systems and Computing*, 165–176. doi:10.1007/978-981-15-3383-9_15

DeVries, T., & Ramachandram, D. (2017). *Skin Lesion Classification Using Deep Multi-Scale Convolutional Neural Networks*. https://arxiv.org/abs/1703.01402

Gordon, R. (2013). Skin cancer: An overview of epidemiology and risk factors. *Seminars in Oncology Nursing*, 29(3), 160–169. doi:10.1016/j.soncn.2013.06.002 PMID:23958214

Harley, A. W. (2015). An interactive node-link visualization of Convolutional Neural Networks. *Advances in Visual Computing*, 867–877. doi:10.1007/978-3-319-27857-5_77

ISIC Archive. (n.d.). *Archive*. ISIC. https://isic-archive.com/

Jaworek-Korjakowska, J., Kleczek, P., & Gorgon, M. (2019). Melanoma thickness prediction based on convolutional neural network with VGG-19 model transfer learning. In *2019 IEEE/CVF Conference on Computer Vision and Pattern Recognition Workshops (CVPRW)*. IEEE. 10.1109/CVPRW.2019.00333

Jeyakumar, J. P., Jude, A., Priya Henry, A. G., & Hemanth, J. (2022). Comparative analysis of melanoma classification using Deep Learning techniques on dermoscopy images. *Electronics (Basel)*, *11*(18), 2918. doi:10.3390/electronics11182918

Mane, S., & Shinde, S. (2018). A method for melanoma skin cancer detection USING Dermoscopy Images. In *2018 Fourth International Conference on Computing Communication Control and Automation (ICCUBEA)*. IEEE. 10.1109/ICCUBEA.2018.8697804

Marks, R. (1995). An overview of skin cancers. *Cancer*, *75*(S2), 607–612. doi:10.1002/1097-0142(19950115)75:2+<607::AID-CNCR2820751402>3.0.CO;2-8 PMID:7804986

Mind, D. S.-D. (n.d.). *Umělá inteligence, machine learning a neuronové sítě*. Data Science - Data Mind. http://www.datamind.cz/cz/Sluzby-Data-Science/umela-intelingence-AI-ML-machine-learning-neural-net

Nasr-Esfahani, E., Samavi, S., Karimi, N., Soroushmehr, S. M. R., Jafari, M. H., Ward, K., & Najarian, K. (2016). Melanoma Detection by Analysis of Clinical Images Using Convolutional Neural Network. In *Proceedings of the 2016 38th Annual International Conference of the IEEE Engineering in Medicine and Biology Society (EMBC)*, (pp. 1373). IEEE. 10.1109/EMBC.2016.7590963

Prabhu. (2019, November 21). Understanding of convolutional neural network (CNN) - deep learning. *Medium*. https://medium.com/@RaghavPrabhu/understanding-of-convolutional-neural-network-cnn-deep-learning-99760835f148

Pujara, A. (2020, July 15). Image classification with MobileNet. *Medium*. https://medium.com/analytics-vidhya/image-classification-with-mobilenet-cc6fbb2cd470

Razzak, I., Shoukat, G., Naz, S., & Khan, T. M. (2020). Skin lesion analysis toward accurate detection of melanoma using multistage fully connected residual network. In *2020 International Joint Conference on Neural Networks (IJCNN)*. IEEE. 10.1109/IJCNN48605.2020.9206881

Ren, Y., Shi, B., Hou, R., Grimm, L., Mazurowski, M. A., Marks, J., King, L., Maley, C., Hwang, S., & Lo, J. (2018). Learning better deep features for the prediction of occult invasive disease in ductal carcinoma in situ through transfer learning. In *Medical Imaging 2018: Computer-Aided Diagnosis*. IEEE. doi:10.1117/12.2293594

PS Staff. (2016). *What Is the Difference between Melanoma And non-Melanoma Skin Cancer?* PSS. https://www.premiersurgical.com/01/whats-the-difference-between-melanoma-and-non-melanoma-skin-cancer/

Xu, J. (2017, August 15). An intuitive guide to deep network architectures. *Medium*. https://towardsdatascience.com/an-intuitive-guide-to-deep-network-architectures-65fdc477db41

Zeiler, M. D., & Fergus, R. (2013). *Visualizing and Understanding Convolutional Networks*. https://arxiv.org/abs/1311.2901

Zheng, Y., Yang, C., & Merkulov, A. (2018). Breast cancer screening using convolutional neural network and follow-up Digital Mammography. *Computational Imaging, III*, 4. doi:10.1117/12.2304564

Chapter 8

Critical Review Analysis on Deep Learning–Based Segmentation Techniques for Water–Body Extraction

Swati Gautam
Maulana Azad National Institute of Technology, India

Ajay Sharma
https://orcid.org/0000-0001-7951-9371
VIT Bhopal University, India

Bhavana Prakash Shrivastava
Maulana Azad National Institute of Technology, India

ABSTRACT

The rapid advancement in the applications of remote sensing imagery had attracted considerable attention from researchers for digital image analysis. Researchers had performed the surveying and delineation of water bodies with excellent efforts and algorithms in the past, but they faced many challenges due to the varying characteristics of water such as its shape, size, and flow. Traditional methods employed for water body segmentation posed certain limitations in terms of accuracy, reliability, and robustness. Rapid growth in the automation category allowed researchers to incorporate deep learning models into the segmentation analysis. Deep learning segmentation models for water body feature extraction have shown promising results based on accuracy and precision. This chapter presents a brief review on the deep learning models used for water-body extraction with their merits over the traditional approaches. It also discusses existing results with challenges faced and future scope.

DOI: 10.4018/978-1-6684-7659-8.ch008

1. INTRODUCTION

Remote sensing technology: a key technique to extract information on land covers and environmental resources. Remotely sensed satellite images and data comprises of spectral, spatial and temporal resolutions. Satellites provide multispectral and hyper spectral images for digital image analysis. These images can be used for studying various image applications such as: image enhancement, image fusion, image segmentation etc. Out of these image segmentation has achieved researcher's interest using remote sensing image analysis. Image segmentation could be applied to variety of fields such as text recognition, land cover mapping, water-body feature extraction etc. Water-body segmentation is an important application of satellite image analysis which separates water and non-water regions. Water (in soil, vegetation or water-bodies) absorbs radiation at NIR bands achieving strong absorption at about 1.4, 1.9 and 2.7 µm. Almost all incident near-infrared and middle infrared (740-2500nm) radiation entering a pure water body is absorbed with negligible scattering. This makes the water-body appearing so dark in the black and white and color-infrared films. Studies have shown that the Landsat series provide effective images on water bodies for particularly carrying out surface water mapping and water delineation.

Rapid urbanization is increasingly deteriorating the water present in the urban environments. Hence, extraction of information from water bodies is essential for continuous evaluation of the status of water bodies. Extracting water-body features with the help of field surveys and GPS monitoring systems was time-consuming and inaccurate. Remote sensing technology has seen tremendous growth in last few years. With the help of this technique, researchers have tried to eradicate the problem of high-cost, more time-consumption and less accuracy as much as possible.

Before coming on to the era of machine learning and deep learning algorithms, researchers used several traditional methods for segmenting the water body images. Single-band, double-band, water-indices or band-ratio methods (NDWI [McFeeters, 1996], MNDWI [Xu, 2006], AWEI [Feyisa et al., 2014], etc.) were initially applied to segment the water bodies. Otsu applied thresholding method (Nobuyuki, 1979) to select the optimum threshold value for water-bodies and provide accurate separation of water-bodies from building shadows. Model-based methods achieved success in segmenting the rough as well as smooth surface water bodies, but these models were more sensitive towards the environmental noise. Though, object-based analysis received increased attention over many years on deriving image objects with high spatial resolutions, but they suffered from low computational efficiency. The need for automation gave rise to the use of artificial intelligence algorithms for water-body segmentation. Deep learning (Du, Cai, Wang et al, 2016) has achieved tremendous popularity in semantic segmentation methodology. The models designed for segmentation are partially or completely based on basic deep learning architectures

(Artificial Neural Network (ANN), Convolutional Neural Network (CNN), and Recurrent Neural Network (RNN)).

The objective of this chapter is to present a brief review on the deep learning architectural models developed for water-body segmentation so far to establish a bridge of research gaps faced in past. The importance of deep learning models and how well these models suite the segmentation process for high resolution remote sensing images had been depicted with its future scope in the field of water-body feature extraction and segmentation. A brief review on existing methodologies has been described with their merits and limitations in Section 2. Section 3 depicts the tabular description of publicly available benchmarking datasets. A qualitative and quantitative comparison on existing results has been described in Section 4 to study the efficacy of the existing models. Finally, the discussion on challenges faced by researchers and conclusion with future scope have been presented in Section 5 and Section 6 respectively.

2. EXISTING METHODOLOGIES

In the recent years, automation methods have achieved great popularity. Deep learning is basically the subset of artificial intelligence. Deep learning has proved to be successful in saving time and learning larger datasets. Unlike traditional methods, deep learning possesses strong learning ability and its practicability makes it more popular among the researchers to explore its applications in accurate segmentation by overcoming the challenges existing with water-bodies. McCulloch proposed the McCulloch-Pitts (MP) model based on the characteristics of neurons. Deep learning came into existence with the development of the Artificial Neural Network. Various deep learning models developed by researchers for water-body segmentation have been discussed further.

2.1 Artificial Neural Network-Based Segmentation Methods

Artificial Neural Networks (ANNs) are developed on the basis of human nervous system where each neuron is assumed to be as functional as a biological neuron. ANNs are able to produce the classification results with higher accuracy due to their non-parametric nature. Multi-Layer Perceptron (MLP) model as shown in Figure 1. Komeil et al. (2015) serves as the basic feed-forward artificial neural network. The three basic layers of the model include: input layer, hidden layer and output layer. Owing to the design of the model; input layer size is equivalent to number of features on which the classification is based. However, the output layer size corresponds

Figure 1. Neural network model used by Komeil et al. (2015)

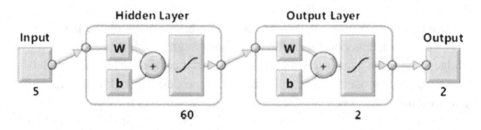

to the number of output classes. So, ultimately, the adjustments are made to the number of hidden layers.

Table 1 summarizes the merits and demerits of some of the ANN-based models utilized for water-body feature extractions.

2.2 Convolutional Neural Network-Based Segmentation Methods

Currently, convolutional neural networks have proved to be efficient for both natural scene and remote sensing images (Cheng et al., 2017). Convolutional layer, pooling layer and fully-connected layer are the three basic layers of any CNN architecture as shown in Figure 2. Feature extraction is done by convolutional layer and pooling

Table 1. Summarization of ANN-based water-body segmentation methodologies

ANN Methodology	Study Site	Merits	Demerits
Texture-based surface water detector along with neural network classifier (Solbo et al., 2003)	Heimdalen area, Norway	Accurately classifies water pixels at low incidence angles.	Could not evaluate the accuracy of water contour.
Classification based Perceptron model to define feature vectors (Kshitij & Prasad, 2015)	Hyderabad city, India	Isolates large-sized water bodies with high classification accuracy.	Unable to extract small water bodies.
Supervised back propagation neural network algorithm (Nilanchal & Rohit, 2014)	Ranchi, Jharkhand	Separates water from land and other built-up features.	Model could not be employed to multi-seasonal datasets.
Neural Network classifier based on non-linear feed forward model with standard back-propagation (Komeil et al., 2015)	Lake Urmia, Iran	Produces high-resolution multi-spectral image and monitors surface water changes of lake.	Multi-temporal image fusion is not suitable for urban water detection.
Based on neural network classification (Pham-Duc, & Prigent,, 2017)	Mekong Delta in Vietnam	Monitoring and quantification of surface water is possible in all weather conditions.	This frameworks requires equalized dataset for better accurate results. Weak in identifying water pixels.

Figure 2. Basic CNN architecture

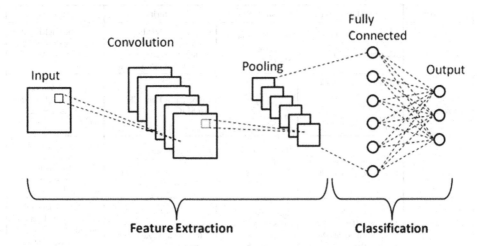

layer does the feature mapping. The use of the fully connected layer is completely dependent on the application.

Table 2 summarizes the merits and the demerits of the water-body segmentation models based on CNN.

2.3 Recurrent Convolutional Network

Recurrent Neural Networks (RNNs) are basically applicable to sequential data analysis such as videos, time-series, continuous text analysis. Recurrent neural networks have been employed to identify the crop temporal pattern in agricultural regions. This LSTM-RNN model (Ziheng et al., 2019) was trained and tested over Cropland Data Layer Dataset. This model provides an overall accuracy of about 97% for 5 classes in the region. Though, RNN proves to be the promising method for analyzing the land scene images in time-series, but, RNN methodology had not been used by the researchers in the field of water-body segmentation yet. Hence, this section needs to be explored more for water-body image analysis.

3. BENCHMARKING DATABASES

Satellite data with very high resolution are defined by a spatial resolution of about 1m. Extensive selection of the high resolution and very high resolution satellite imagery had been commercially made available by various leading providers under public domain and private domain. The datasets presented in Table 3 have been made

Table 2. Summary of CNN-based water-body segmentation methodologies

Network	CNN Methodology	Study Site	Merits	Demerits
Fully Convolutional Network	CNN based extraction model with logical regression classifier (Long et al., 2017)	XinJiang (river), Hubei (reservoir), Shenyang (pond)	Outperforms SVM and ANN by extracting water features from dense mountainous environment.	Causes redundancy due to the use of segmented image patches for fusion.
	CNN based model for performing segmentation between sea and land with in-dock ship detection (Lin et al., 2017)	Images from Google maps and GaoFen-1 dataset at different coordinates.	Outperforms DenseCRF and SLIC.	Ship target detection on water is only limited to navy ships and Oil tanks.
	DeepWaterMap based on FCN (Furkan et al., 2017)	High altitudes of North America WRS-2 Path/Row=49/24	Outperforms simple MNDWI and ANN by discriminating water from other classes.	False positives obtained on underpresented classes leading to confusion with shadows and clouds.
Multi-scale Network	CNN model with MSR-Net+ EA (erasing attention) module (Duan & Hu, 2019)	Rural areas collected as images from GID and Digital Globe.	Outperforms FCN, SegNet, ExFuse and DeepLabv3+ with more accurate results for narrow rivers.	Baseline model with 50 layers could not provide efficient results for water segmentation maps.
	OTOP (MSCNN combined with Google Earth Engine) (Yudie et al., 2019)	Urban areas of Central China (Wuhan).	Outperforms MNDWI and Random Forest Method in detection of urban water bodies.	Unable to detect water fineness when compared with pixel-based extraction.
	MWEN (Multi-scale Water Extraction) model (Guo et al., 2020)	Beijing-Tianjin-Hebei region with agricultural water areas and wetland.	Accurately extracts open ponds and plateau lakes with different sizes and suppresses building noise and highways.	Unable to suppress the noise from sports field which affects the classification accuracy.
Encoder-Decoder Network	Deep CNN model for transfer learning based on SegNet (Mina & Wathiq, 2018)	City are of Dubai.	Improves boundary delineation.	Provides false positive in detecting the water pixels in some cases.
	Deep U-Net-CRF: a modified FCN (Feng et al., 2019)	Images from GF-2 and WV-2 from China.	Use of CRF improves water body extraction quality with the improvement in accuracy.	Due to the longer training times the computation cost is high.
Others	Self-adaptive Pooling Convolutional Neural Network (SAPCNN) model (Chen, Fan, & Yang, 2018)	Images of Beijing and Tianjin from ZY-3 and GF-2.	Avoids over-segmentation in homogeneous sub-regions, removes blurry boundaries, identifies small water region more accurately.	The pooling factor 'u' needs to be properly optimized for the efficient reduction in the removal of image feature.
	Restricted Receptive Field Deconvolutional Network (RRFDeConvNet) (Miao et al., 2018)	Google earth images from Suzhou and Wuhan, China.	Improved performance in case of blurry boundaries and enhances visual representation.	EW Loss function does not provide any improvement from mathematical point of view.
	CNN based framework for global identification of reservoirs (Fang et al., 2019)	Reservoirs at global scale.	First attempt at global scale to recognize lakes and reservoirs.	Seasonal Fluctuations affect the determination of lakes around roads and highways.
	Mask R-CNN (Noppitak et al., 2020)	North-eastern region of Thailand.	Reduced network learning time. Identifies natural and artificial water bodies.	Model can only evaluate RGB color water body images.

publicly available for the future researchers to experiment their algorithm for the segmentation/feature extraction of water bodies and analysis of the results obtained.

4. COMPARATIVE RESULT ANALYSIS

Image Segmentation algorithms need to be evaluated on various indices to validate their efficiency and effectiveness and compared with other results. Hence, this section is bifurcated into two sub-sections with first sub-section describing the evaluation parameters and the second showing comparison of results.

Table 3. Description on some publicly available databases for water-body feature extraction

S.No.	Dataset/Satellite Detail	Classes	Image Type	Resolution	Download Link	References
1.	Sentinel-2 Level-1C product Database	Open water resources, buildings and vegetation	RGB image	10-20m	http://sci.hub. copernicus.eu/	(Du, Zhang, & Ling, 2016)
2.	Landsat 8 Satellite imagery	Lakes & Reservoirs	RGB image	30m	http://glovis.usgs. gov	(Anoj et al., 2018; Sanchez et al., 2018)
3.	GLS2000 Dataset (Landsat 7 ETM+ images)	Inland surface water bodies	RGB image	30m	http://www. landcover.org (GIW version 1.0)	(Feng et al., 2016)
4.	NWPU-RESISC45	700 images per class with 45 scene classes including water bodies	RGB image	~30m-0.2m	http://www. escience.cn	(Cheng et al., 2017)
5.	Google Earth images Database	Lakes, reservoirs, rivers, ponds, paddies and ditches.	RGB image	0.5m	http://earth. google.com/	(Miao et al., 2018)
6.	GaoFen-2 (GF-2) satellite images Database	Lakes, rivers and ponds.	RGB image	0.8m	http://www. DigitalGlobe. com/	(Feng et al., 2019)
7.	ALOS/AVNIR-2 satellite images Database	Water body images	RGB/ NIR image	10m (4 bands)	https://asf.alaska. edu/data-sets	(Ya'nan et al., 2014)
8.	Landcover Classification DeepGlobe Dataset	803 images of urban, agriculture, rangeland, forest, water & unknown classes.	RGB images	4m	http://www. kaggle.com	(Duan & Hu, 2019)

4.1 Evaluation Indices

Some of the standard metrics (accepted by the scientific community) have been described and used by the researchers for the assessment of segmentation methods. Table 5 describes the accuracy indices (Anoj et al., 2018; Yang et al., 2018) used for evaluation along with their definition and formulae.

Where,

True Positive (TP) = number of truly extracted target pixels, True Negative (TN) = number of truly rejected non-target pixels, False Positive (FP) = number of falsely extracted target pixels, False Negative (FN) = number of undetected target pixels, T= total number of pixels in accuracy assessment and Σ = (TP+FP)(TP+FN) + (FN+TN)(FP+TN) is the chance accuracy.

Other indices used for evaluation have been described in Table 5.

Where, $|A|$ and $|B|$ are the cardinal of set A and set B, respectively.

Some of the techniques could also be validated using water edge pixel parameters that include Edge Overall Accuracy (EOA), Edge Omission Error (EOE) and Edge Commission Error (ECE) (Chen, Fan, & Yang, 2018).

Based on these performance indices the results for the various methodologies have been discussed in the next segment.

Table 4. Accuracy indices for evaluation

Evaluation Index	Definition	Formulae	Range
Producer's Accuracy (PA)/Recall/Sensitivity	Test's ability to correctly detect the target pixels in an image.	$PA = \dfrac{TP}{TP + FN}$	0 to 1
User's Accuracy (UA)/ Precision	Pixel probability being classified into the target object.	$UA = \dfrac{TP}{TP + FP}$	0 to 1
Overall Accuracy (OA)	Measurement of how accurate a particular technique is.	$OA = \dfrac{TP + TN}{T}$	0 to 1
Kappa Coefficient	Like, overall accuracy, describes the efficacy of technique.	Kappa Coefficient = $\dfrac{T(TP + TN) - \Sigma}{T^2 - \Sigma}$	-1 to +1

Table 5. Some other evaluation parameters

Evaluation Index	Definition	Formulae
Jaccard Index or (Intersection over Union) IoU (Muhadi et al., 1825)	Validates semantic segmentation results and expresses the number of objects two sets have in common.	Jaccard Index $= \dfrac{\lvert A \cap B \rvert}{\lvert A \cup B \rvert}$
Dice Similarity Coefficient (DSC)/F1-Score	Spatial overlap index ranging between 0% to 100%, with 0% indicating the no spatial overlap between two segmentation result and 100% indicating complete overlap	DSC/F1-Score $= \dfrac{2\lvert A \cap B \rvert}{\left(\lvert A \rvert + \lvert B \rvert\right)}$

4.2 Comparison of Results

Water-body image segmentation analysis in itself is a very tedious task and offers certain challenges to the researchers due to the presence of noise and other artefacts. The qualitative analysis of the algorithms is always complemented with the quantitative analysis to provide better information regarding the efficiency of a particular method. Though, it is difficult to compare the methods on a generalized basis because of their use in different study sites and different databases for various applications. But, the methods can be quantitatively compared for some test sites to provide the researchers with a comprehensive idea about the methods to be used in certain applications of water body feature extraction.

In this section, a comparison on results of certain state-of-the-art deep learning methodologies employed in water-body segmentation has been provided to help the researchers with the idea of presenting the basic overview of the techniques along with their performance on studied test sites.

In Table 6, Multi-scale Refinement Network (MSR-Net) (Duan & Hu, 2019) is compared with other four approaches based on deep learning (FCN (Shelhamer et al., 2017), SegNet (Badrinarayanan et al., 2017), Ex-Fuse (Zhang et al., 2018), and DeepLabv3+ (Chen, Zhu, Papandreou et al, 2018)). For their fair comparison, these models were implemented with the setting similar to the MSR-Net. MSR-Net refines the segmentation map by embedding multi-scale image features. This multi-scale refinement strategy (MSR-Net) outperforms the considered competing methods in four evaluation indices (IoU, Precision, F1 and Kappa), as this network is successful in generating the unbroken segmentation maps of linear rivers.

Convolutional Neural Networks (CNNs) have achieved better response from the researchers due to their outstanding performances in comparison to the artificial neural networks.

Table 6. Quantitative comparison of the existing methods on DeepGlobe dataset

S. No.	Methods	IoU (%)	Precision (%)	Recall (%)	F1 (%)	Kappa
1.	FCN (Shelhamer et al., 2017)	84.37	88.49	94.78	91.52	0.896
2.	SegNet (Badrinarayanan et al., 2017)	77.42	78.72	97.91	87.27	0.847
3.	Ex-Fuse (Zhang et al., 2018)	84.78	88.63	95.12	91.76	0.899
4.	DeepLabv3+ (Chen, Zhu, Papandreou et al, 2018)	85.19	88.72	95.53	92.00	0.902
5.	MSR-Net (Duan & Hu, 2019)	86.20	89.65	95.73	92.59	0.909

However, CNN models often lead to the generation of blurry boundaries when used alone. Hence, their performance is modified with the use of certain types of pooling combinations with the CNN models and the accuracy in terms of edges is determined by the evaluation of three parameters (EOA, EOE, and ECE) and it could be identified from Table 7 that the Self adaptive pooling Network in combination with CNN (Chen, Fan, & Yang, 2018) provides higher edge overall accuracy and minimum omission and commission error values in comparison to other pooling approaches.

After the successful development in CNN architectural models, the researchers decided to increase the hidden layers in the convolutional neural networks to increase the applicability of the models to various areas. This gave rise to the deep convolutional neural networks. Table 8 describes the deep convolutional neural networks in combination with three different approaches and their efficiency is measured in terms of accuracy. It is visible from the tabular description that the restricted receptive field deconvolutional network (DeconvNet-RRF) (Miao et al., 2018) outperforms the two approaches i.e. DeconvNet-VGG and DeconvNet-ResNet18 by giving the higher accuracy value of 0.965. This is because RRF DeconvNet compresses the redundant layers in comparison to the other two and it no longer relies on a pre-trained model. Hence, gives better results in comparison to other models.

Table 7. Water-Edge pixel parameter analysis using different CNN architectural models (Chen, Fan, & Yang, 2018)

S. No.	Parameters	Self-Adaptive Pooling + CNN	Max Pooling + CNN	Average Pooling + CNN
1.	Edge Overall Accuracy (EOA)	97.82	94.21	91.27
2.	Edge Omission Error (EOE)	0.94	2.63	6.24
3.	Edge Commission Error (ECE)	1.24	3.16	2.49

Table 8. Accuracy comparison for different deep convolutional neural networks (Miao et al., 2018)

S. No.	Method	Accuracy
1.	DeconvNet-VGG	0.962
2.	DeconvNet-ResNet18	0.96
3.	DeconvNet-RRF	0.965

5. DISCUSSION

This study on the review of deep learning architectural models for water-body segmentation had come up with the identification of four major findings that serve as challenges to the researchers.

- **Lack of data with ground-truth**: Unavailability of masks and ground-truth in the archieve files makes difficult for the researchers to compare their methods with the existing methodologies. Hence, there is a need to develop the datasets with their masks and labels to make the comparison of methods easy.

- **Need to develop models at global scale**: As there are variety of resources existing in the environment with different variabilities hence developing different model for each type would be difficult, time-consuming and expensive. Hence, researchers need to explore more in the field of developing models with global feature extraction.

- **Demand for fineness and pixel-level accuracy:** In water-body segmentation, water-bodies need to be segmented with high accuracy by separating its boundaries from land and other connected objects to achieve complete flow of water bodies at either sides and attain higher classification accuracy.

- **Need to extract smaller water regions:** Urbanization and social development has led to the vanishment of certain small water bodies that were the source of life for many living habitats anf creatures. Hence, there is a need to develop model to extract the sub-urban smaller water bodies in the dense urban environment so that they may be identified and saved for further sustainable management.

6. CONCLUSION AND FUTURE SCOPE

The aim of this review was to develop a brief understanding of existing deep learning methods in the field of water-body segmentation and make the upcoming researchers aware of the challenges faced by the researchers in the past. Information on publicly available datasets suitable for water-body segmentation enhances the quality of study which tries to subdue the major challenge being faced by the current researchers. The description on the evaluation parameters and the comparative result analysis of some of the state-of-the-art methods enables the reader understand the limitations of various algorithms and find out the direction of their research. Finally an attempt has been made to point out the findings from the methods explored so far with a brief discussion on each finding. These discussions will enable the researchers to develop the algorithm with universal acceptance and higher classification accuracy.

For future applications of deep learning models, the flaws observed in previously used methods need to be eliminated for better research results. Based on the methods surveyed for this study, certain recommendations can be listed for future research. They include:

- **Development of large-sized datasets:** Previous publications have experienced the unavailability of very large datasets ready for deep learning models. Hence, efforts can be done by the future researchers in developing the large sized deep learning ready datasets for performing comparative research study.
- **Need for Interpretable Deep Models:** No doubt, DLmodels have provided excellent results on benchmarking datasets but there are still some questions that have not been answered yet. These questions include: What the deep learning model is actual learning? How to interpret the features learned by the DL-model? What is the minimum amount of data needed to train a DL-model to achieve significant amount of accuracy? All these questions need to be answered for better understanding and development of deep learning models.
- **Universal Models:** There is a need to develop the model that can extract the water-body features at global scale. One such attempt has been made in the past but the application of model is only limited to the identification of reservoirs and lakes. Further, attempts can be made in future for extracting out features from all types of water resources using single deep learning model.

The application of the advanced deep learning techniques into the remote sensing imagery would lead to vast opportunities to propel more advances in water sciences due to the increasing development in automation and artificial intelligence.

REFERENCES

Anoj, S., Tri Dev, A., & Ha, L. (2018). Evaluation of Water Indices for Surface Water Extraction in a Landsat 8 Scene of Nepal. *Sensors, 18*, 1-15.

Badrinarayanan, V., Kendall, A., & Cipolla, R. (2017). SegNet: A deep convolutional encoder-decoder architecture for image segmentation. *IEEE Transactions on Pattern Analysis and Machine Intelligence*, *39*(12), 2481–2495. doi:10.1109/TPAMI.2016.2644615 PMID:28060704

Chen, L. C., Zhu, Y., Papandreou, G., Schroff, F., & Adam, H. (2018). Encoder decoder with Atrous separable convolution for semantic image segmentation. *Proc. Eur. Conf. Comput. Vis. (ECCV),* (pp. 801–818). IEEE. 10.1007/978-3-030-01234-2_49

Chen, Y., Fan, R., & Yang, X. (2018). Extraction of Urban Water Bodies from High-Resolution Remote-Sensing Imagery Using Deep Learning. *Water, 10*, 585.

Cheng, G., Han, J., & Lu, X. (2017). Remote sensing image scene classification: Benchmark and state of the art. *Proceedings of the IEEE*, *105*(10), 1865–1883. doi:10.1109/JPROC.2017.2675998

Du, X., Cai, Y., Wang, S., & Zhang, L. (2016). Overview of deep learning. *31st Youth Academic Annual Conference of Chinese Association of automation (YAC),* (pp.159–164). YAC. 10.1109/YAC.2016.7804882

Du, Y., Zhang, Y., Ling, F., Wang, Q., Li, W., & Li, X. (2016). 'Water Bodies' Mapping from Sentinel-2 Imagery with Modified Normalized Difference Water Index at 10-m Spatial Resolution Produced by Sharpening the SWIR Band. *Remote Sensing, 8*(354), 1-19.

Duan, L., & Hu, X. (2019). Multiscale Refinement Network for Water-body Segmentation in High-Resolution Satellite Imagery. *IEEE Geoscience and Remote Sensing Letters*, *17*(4), 686–690. doi:10.1109/LGRS.2019.2926412

Fang, W., Wang, C., Chen, X., Wan, W., Li, H., Zhu, S., Fang, Y., Liu, B., & Hong, Y. (2019). Recognizing Global Reservoirs from Landsat 8 Images: A Deep Learning Approach. *IEEE Journal of Selected Topics in Applied Earth Observations and Remote Sensing*, *12*(9), 3168–3177. doi:10.1109/JSTARS.2019.2929601

Feng, M., Sexton, J. O., Channan, S., & Townshend, J. R. (2016). A global, high resolution (30-m) inland water body dataset for 2000: First results of a topographic–spectral classification algorithm. *International Journal of Digital Earth*, *9*(2), 113–133. doi:10.1080/17538947.2015.1026420

Feng, W., Sui, H., Huang, W., Chuan, X., & Kaiquiang, A. (2019). Water body Extraction from Very High Resolution Remote Sensing Imagery using Deep U-Net and Superpixel Based Conditional Random Field Model. *IEEE Geoscience and Remote Sensing Letters*, *16*(4), 618–622. doi:10.1109/LGRS.2018.2879492

Feyisa, G. L., Meilby, H., Fensholt, R., & Proud, S. R. (2014). Automated water extraction index: A new technique for surface water mapping using Landsat imagery. *Remote Sensing of Environment*, *140*, 23–35. doi:10.1016/j.rse.2013.08.029

Furkan, I., Bovik, A. C., & Paola, P. (2017). Surface Water mapping by Deep Learning. *IEEE Journal of Selected Topics in Applied Earth Observations and Remote Sensing*, *10*(11), 4909–4918. doi:10.1109/JSTARS.2017.2735443

Guo, H., He, G., Jiang, W., Yin, R., Yan, L., & Leng, W. (2020). A Multi-Scale Water Extraction Convolutional Neural Network (MWEN) Method for GaoFen-1 Remote Sensing Images. *ISPRS International Journal of Geo-Information*, *9*(4), 189. doi:10.3390/ijgi9040189

Komeil, R., Anuar, A., Karim, S., & Sharifeh, H. (2015). A new approach for surface water change detection: Integration of pixel level image fusion and image classification techniques. *International Journal of Applied Earth Observation and Geoinformation*, *34*, 226–234. doi:10.1016/j.jag.2014.08.014

Kshitij, M., & Prasad, R. (2015). Automatic Extraction of Water Bodies from Landsat Imagery Using Perceptron Model. *Journal of Computational Environmental Sciences*. . doi:10.1155/2015/903465

Lin, H., Shi, Z., & Zou, Z. (2017). ʻMaritime Semantic Labeling of Optical Remote Sensing images with multiscale fully convolutional networks. *Remote Sensing (Basel)*, *9*(5), 480. doi:10.3390/rs9050480

Long, Y., Wang, Z., Tian, S., Ye, F., Ding, J., & Jun, K. (2017). Convolutional Neural Networks for Water-body Extraction from Landsat Imagery. *International Journal of Computational Intelligence and Applications*, *16*(1).

McFeeters, S. K. (1996). The use of the normalized difference water index (NDWI) in the delineation of open water features. *International Journal of Remote Sensing*, *17*(7), 1425–1432. doi:10.1080/01431169608948714

Miao, Z., Kun, F., Hao, S., Xian, S., & Yan, M. (2018). Automatic Water-body Segmentation From High Resolution Setellite Images via Deep Networks. *IEEE Geoscience and Remote Sensing Letters*, *15*(4), 602–606. doi:10.1109/LGRS.2018.2794545

Mina, T., & Wathiq, M. (2018). Detection of Water-Bodies Using Semantic Segmentation. *International Conference on Signal Processing and Information Security (ICSPIS)*. IEEE.

Muhadi, N. A., Abdullah, A. F., Bejo, S. K., Mahadi, M. R., & Mijic, A. (1825). 2020, 'Image Segmentation Methods for Flood Monitoring System', Water. *An MDPI Journal, 12*, 1–10.

Nilanchal, PRohit, M. (2014). Extraction of Impervious Features from Spectral indices using Artificial Neural Network. *Arabian Journal of Geosciences, 8*(6).

Nobuyuki, O. (1979). A Threshold Selection Method from Gray-Level Histograms. *IEEE Transactions on Systems, Man, and Cybernetics, SMC-9*(1), 62–66.

Noppitak, S., Gonwirat, S., & Surinta, O. (2020). *Instance Segmentation of Water-body from Aerial Image using Mask Region based Convolutional Neural Network.* The 3rd International Conference on Information Science and Systems (ICISS), Cambridge, UK.

Pham-Duc, B., Prigent, C., & Aires, F. (2017). Surface water monitoring within Cambodia and the Vietnamese Mekong Delta over a year, with Sentinel-1 SAR observations. *Water, 9*(6), 366

Sanchez, G., Dalmau, O., & Alarcon, T. (2018). Selection and Fusion of Spectral Indices to Improve Water Body Discrimination. *IEEE Access, 6,* 72952-72961.

Shelhamer, E., Long, J., & Darrell, T. (2017). Fully convolutional networks for semantic segmentation. *IEEE Transactions on Pattern Analysis and Machine Intelligence, 39*(4), 640–651. doi:10.1109/TPAMI.2016.2572683 PMID:27244717

Solbo, S., Malnes, E., Guneriussen, T., Solheim, I., & Eltoft, T. (2003). Mapping surface-water with Radarsat at arbitrary incidence angles. *Geoscience and Remote Sensing Symposium, IGARSS'03 Proceedings*. IEEE International. 10.1109/IGARSS.2003.1294494

Xu, H. (2006). Modification of normalised difference water index (MNDWI) to enhance open water features in remotely sensed imagery. *International Journal of Remote Sensing, 27*(14), 3025–3033. doi:10.1080/01431160600589179

Ya'nan, Z., Luo, J., Shen, Z., Hu, X., & Yang, H. (2014). Multiscale Water Body Extraction in Urban Environments from Satellite Images. *IEEE Journal of Selected Topics in Applied Earth Observations and Remote Sensing, 7*(10), 4301–4312. doi:10.1109/JSTARS.2014.2360436

Yang, X., Qin, Q., Grussenmeyer, P., & Koehl, M. (2018). Urban surface water body detection with suppressed built-up noise based on water indices from Sentinel-2 MSI imagery. *Remote Sensing of Environment*, *219*, 259–270. doi:10.1016/j. rse.2018.09.016

Yudie, W., Zhiwei, L., Chao, Z., Xia, G., & Shen, H. (2019). Extracting Urban Water by Combining Deep Learning and Google Earth Engine. *IEEE journal of selected topics in applied earth observations & Remote Sensing, 13*, 768-781.

Zhang, Z., Zhang, X., Peng, C., Xue, X., & Sun, J. (2018). ExFuse: Enhancing feature fusion for semantic segmentation. *Proc. Eur. Conf. Comput. Vis. (ECCV),* (pp. 269–284). IEEE. 10.1007/978-3-030-01249-6_17

Ziheng, S., Liping D., & Hui, F. (2019). Using long short-term memory recurrent neural network in land cover classification on Landsat and Cropland data layer time series. *International Journal of Remote Sensing, 40*(2), 593-614. doi:10.1080/014 31161.2018.1516313

Chapter 9

Hybrid Approaches for Plant Disease Recognition:
A Comprehensive Review

S. Hemalatha
School of Computer Science Engineering and Information Systems, Vellore Institute of Technology, India

Athira P. Shaji
School of Computer Science Engineering and Information Systems, Vellore Institute of Technology, India

ABSTRACT

Plant diseases pose a significant threat to agriculture, leading to yield and quality losses. Traditional manual methods for disease identification are time-consuming and often yield inaccurate results. Automated systems leveraging image processing and machine learning techniques have emerged to improve accuracy and efficiency. Integrating these approaches allows image preprocessing and feature extraction to be combined with machine learning algorithms for pattern recognition and classification. Deep learning, particularly convolutional neural networks (CNNs), has revolutionized computer vision tasks, enabling hierarchical feature extraction. Hybrid methods offer advantages such as improved accuracy, faster identification, cost reduction, and increased agricultural productivity. This survey explores the significance and potential of hybrid approaches in plant disease identification, addressing the growing need for early detection and management in agriculture.

1. INTRODUCTION

Agriculture is one of the biggest contributions to a country's economy. The plant diseases can cause severe negative influence on the quantity and quality of agricultural

DOI: 10.4018/978-1-6684-7659-8.ch009

products. Thus, it is very important to identify them in the early-stage and control the spread over its neighborhood. Early disease detection is essential for getting better yield in the agricultural domain. The lag or failure in identification can lead to low production and loss of economy to the farmers. In the beginning, the detection of plant diseases had been done manually by monitoring plants, collecting the specimen of diseases in plants and consulting the agricultural experts. This process was more time consuming and expensive as the farmers would have to travel a long distance to meet the experts. Moreover, continuous monitoring of plant is cumbersome in case of long plantations (Tucker CC & Chakraborty, 1997).

A solution arrived to solve these issues in terms of image processing approaches (Bock et al.,2010; Lee and Chung,2005). The plant diseases occur mostly due to microorganisms like bacteria, fungus, virus and protozoa. These organisms can turn the plant leaves into yellowish, wilted and molted. Some of the diseases make the leaves with black spot, downy mildew, bacterial built etc. (Mesquita,2011). Thus, it has become possible to apply image acquisition and image processing appropriately to identify damaged leaves in plants (Al-Hiary et al.,2011; Sena et al.,2003). Fundamental Image processing methods such as edge detection and thresholding segmentation (Raji & Alamutu,2005; Zhang et al., 2005) were employed around a decade back for plant disease detection. Furthermore, color image processing (Camargo& Smith, 2009; Bama et al.,2011; Gocławski et al.,2009) techniques were used to identify plant diseases.

With advancements in computer vision and pattern recognition, several studies for plant disease diagnosis have been proposed, with the goal of identifying the diseases through affected leaves. Pattern classification methods helped to identify disease causing agents in plants (Camargo & Smith,2009). It was thought that image processing methods along with machine learning based algorithms could be utilized for the identification and classification of plant diseases. Because the inclusion of machine learning concept for the purpose could help to discriminate appropriate targets when fed with suitable information. Machine learning techniques such as naïve Bayes classifier, Decision tree classifier and perceptron models were used for training the model (Jamuna et al., 2010; Korada et al.,2012; Revathi et al..2011). Various feature selection and rule generation techniques (Phadikar et al..2013) have been proposed and improvised the results. Multi disease identification is done by

proposing a Leaf doctor algorithm (Pethybridge S J&Nelson,2015) that was deployed on a real smartphone application and validated under real field conditions. The computer vision system detects and classifies the crop diseases through image acquisition, feature extraction, feature selection and classification (DeCost BL& Holm; Han et al., in 2015).

The Artificial Bee Colony (ABC) algorithm combined with Negative Selection Algorithm (NSA) (Mishra PK& Bhusry, 2016) has been used to classify iris plant datasets. The combination of ABC with NSA improves the global convergence behavior. For automatic classification of plant diseases, support vector machine (Griffel et al., 2018) found to be more powerful and have reduced the need for human intervention. Depending on the nature of the probability model, machine learning techniques can be either supervised or unsupervised (Griffel et al.,2018; Kothari J D, 2018). Supervised classification is carried out to obtain a class, while in unsupervised approaches, there is no need of training. Some of the important classification techniques includes k-nearest neighbor (KNN), Support vector machine (SVM) and random forest method (RF) (Thanh Noi P & Kappas M.,2017) SVM concepts based on hyperspectral reflectance (Fu et al., 2019) concepts also achieved better results.

Some real time applications have been developed with computer vision and machine learning techniques by certain firms to identify the plant disease identification and classification. A well-known mobile based app named plantix which utilize the computer vision and machine learning technique in crop disease identification and nutrient deficiencies (Plantix). The app identifies the disease through the images sent by farmers by applying Artificial Intelligence. Identification of disease and necessary remedies are suggested by the application. Other well-known apps which use this technique are Agrio, Crops AI, Leaf Doctor, Purdue Tree Doctor and Leaf Plant Tech etc. (Themamapirate). Region proposal network (Guo et al.,2020) can be utilized for disease detection and recognition.

Deep learning bridles the problem arises due to traditional algorithms and it could attract more researcher's attention. It is now widely used in patter recognition, computer vision, speech recognition and natural language processing and so on. Deep learning models (Guo et al.,2020; Chohan et al,2020) became familiar among researchers since they can process huge volume of data and it is cost effective. Deep neural network (DNN) (Cristin et al, 2020; Mishra et al.,2020; Ramesh S, Vydeki,2020) models have been used to investigate the disease in both vegetables, grain and flower plants like maize, sunflower, oats, cotton, wheat, rice etc. Multi classifier cascade method by using SVM and CNN can be used for automatic observation (Hassan et al., 2021).

Since 2012 onwards, Convolutional neural network (Rathore NP& Prasad L,2020; Rahman et al.,2020; Agarwal et al.,2020) based methods have been used for disease symptom detection and recognition. CNN improves classification accuracy and is used for prediction problems. In addition, VGG16 and InceptionV3, Mobile Net, architecture can be used to detect rice diseases.

The classifiers prime aim is to identify whether a plant is affected by a disease or not. The classification is handled through supervised, unsupervised and semi-supervised method (Griffel et al.,2020; Guo et al.,2020; Adams et al.,2020).

2. GENERAL ARCHITECTURE

The general architecture of the plant disease diagnosis and classification system is explicated as shown in Figure 1. The image acquisition, pre-processing, segmentation, feature extraction and classification are all part of the model. The image acquisition entails capturing images with equipment like digital cameras, CCD color camera, android mobile, hyperspectral imaging system etc. The images were taken in either controlled or natural conditions. The performance of the disease detection system depends on the acquired image quality. Image acquisition step is followed by pre-processing stage, which is carried out to make the processing go more smoothly. The following diagram depicts the architecture of the plant disease diagnosis and classification technique.

The pre-processing step followed by segmentation of the image. The segmentation process helps to identify the healthy and affected samples. Common approaches like color-based, edge-based, threshold-based techniques are extensively utilized for segmentation purposes. The spot color-based segmentation approaches are developed based on the color differences in the affected leaf. The properties of an images like color, texture, shape features, contrast, entropy area are utilized in order to extract the features.

The methods like classical gray level co-occurrence matrix (GLCM), spatial gray level dependence matrix (SGDM), Fourier based fractal descriptors (FBFD), speeded-up robust features (SURF), histogram of oriented gradients (HOG), scale-invariant feature transform (SIFT) is some of the methods used for feature extraction purposes.

3. METHODOLOGIES BASED ON NEURAL NETWORKS AND IMAGE PROCESSING

Integration of machine learning in image processing technology had showed very much importance in agriculture field. It is desperately important in complex image processing applications. Number of image processing algorithms can be integrated with some machine learning components to improve adaptation. The use of computer vision technology combined with machine learning technique is beneficial and effective in the recognition of plant diseases. This section

Figure 1. General architecture

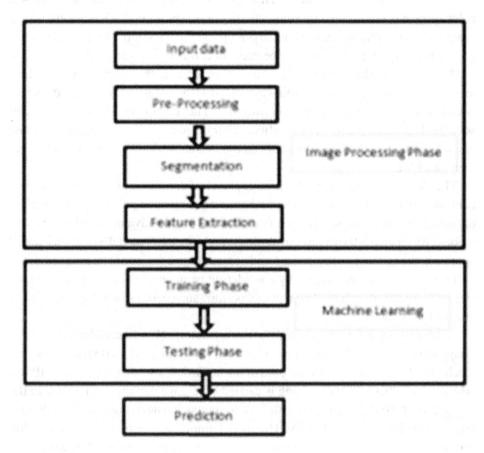

provides an overview of the numerous approaches undertaken to identify and classify plant diseases.

A computer vision based neutrosophic methodology for identifying and classifying leaf diseases proposed by (Dhingra et al., 2019). The segmentation is carried out with neutrosophic logic-based segmentation technique. To detect the condition of the leaves, a new feature subset using texture, color and histogram is evaluated with the help of segmented regions. The features acquired by the segmentation procedure combine the intensity and texture discrimination power of leaves. The methodology employs nine classifiers, and it was observed that the random forest classifier added a superiority over other approaches.

A CNN based technique for recognizing plant leaf diseases was proposed by (Nanehkaran et al.,2020). In this approach, plant disease recognition is carried through segmentation and image classification. The segmentation is carried with hue, saturation and intensity based HIS and LAB based hybrid approach. HIS is a color space it is represented by Hue, Intensity and Saturation. LAB is another color space that could achieved uniformity in perception and it is similar to HIS. Therefore, both these techniques together can be used for achieving image segmentation.

A smart agriculture-based object detection scheme that combines both deep learning and tweaked transfer learning models (Boukhris et al.,2020). Object detection is done using Mask-Region based convolutional neural network concepts and then severity degrees in plants are classified. After detection of specific diseases, they are suggesting some preventive methods too. Visualization is done by contouring their exact locations.

The identification and classification of plant leaf diseases using a depth separable convolution model proposed by (Sk Mahmudul Hassan et.al.,2021) A large number of parameters and a high computational cost are two disadvantages of early CNN models. By replacing the usual convolution layer in the CNN model with the aforesaid approach, the number of parameters and computation costs could be reduced. Final performance evaluation is done by considering parameters such as dropout, batch size and number of epochs. The model's performance was tested using a variety of deep learning models.

A peach disease detection based on the Asymptotic Non-Local Means (ANLM) image algorithm, parallel convolution neural network (PCNN) and extreme learning machine (ELM) optimized with the aid of linear particle swarm optimization (IPSO) (Huang et al.,2020). The denoising is carried to remove the interference due to the complex background in an image with ANLM approach. The characteristics of disease are identified with parallel CNN. The model employed improved ELU activation function. The replaced function traditionally used RELU. Further, SoftMax layer is replaced by linear particle swarm optimized ELM (IPELM).

A leaf classification by using a concept called Bag of Features (BOF) and Dual Output Pulse Coupled Neural Network (DPCNN) is proposed by (Yaonan Zhang et.al, 2020). The DPCNN has features such as translation, rotation, scale invariance and robustness. Also, the feature classification speed can be improved by using uniform sampling method and simple fragments. The proposed BOF_DPCNN method is divided into four stages such as pre-processing, acquisition of DPCNN pulse images, low level feature extraction and feature coding. After feature extraction, SVM classifier is used for classification on Flavia dataset.

Five species of leaves using intelligent computer vision technology. Imaging, pre-processing, feature extraction, discriminant feature selection, hybrid ANN-genetic algorithm and classification employing three different types of classifiers are the main phases in this system (Sajad Sabzi et.al.,2020). Three classifiers exploited in this study includes hybrid Artificial neural network-biogeography based optimization (ANN-BBO), classifier, hybrid ANN-Anti bee colony and LDA. The gray level co-occurrence matrix (GLCM) is used to extract texture features such as homogeneity, entropy, smoothness, third moment, average and standard deviation. There were 108 texture features and 29 shape characteristics extracted in total. Color features extracted in this study is divided into statistical color features and vegetation index color features. Thus, LDA classifier used here could achieve an accuracy of about 93.99%. From the overall observations it could be noticed that LDA classifier is superior to both ANN-ABC and ANN-BBO.

A spike detection algorithm using image processing techniques achieved better results (Narendra Narisetti et al.,2020). For classification of spike and non-spike pixels neural network is used. Wavelet decomposition is significant because it aids in the creation of non-redundant compressed image, which reduces computation complexity. In addition, the wavelet coefficient image utilized for texture characterization and generation of binary plant images is done by binarization technique. The suggested method is feasible for high throughput analysis of HTP detection. Thus, the above method is expected to perform well in germplasm with low biomass and tillering.

An optimized deep neural network for the paddy plant leaves classification showed improved results (Ramesh and Vydeki,2020). The images were taken directly from the field with a camera. The pre-processing of images is done by removing background of RGB images. RGB images are transformed to HSV color space for background removal, and the hue and saturation components of binary images are extracted to split the diseased and non-diseased parts during pre-processing. K-means clustering approach is used for segmentation. The k means clustering is handled on the hue part of HSV model. The classification is done using a DNN that has been optimized using a java optimization algorithm. A feedback loop is established in order to achieve stability.

Five major diseases caused in eggplant by pests and pathogens identified(Aravind and Raja,2020). Overfitting issues were found in deep learning models trained with a smaller number of datasets. Therefore, for increasing the number of sample images, augmentation of datasets was performed. The performance was put to the test in five trials. For training and validation, a pre-trained deep learning models called VGG16 was employed. The available dataset is also compared to the most prominent deep learning architectures. Grayscale images had the lowest classification accuracy with Alexnet among the other color spaces.

A framework for the classification of medicinal plant leaf based on multispectral and texture features using machine learning approach (Samreen et al.,2021). A computer vision laboratory setup was setup to collect multispectral and digital image datasets which helped to offer datasets that were free of noise. Using feature selection, they were able to identify the most important features and eliminate the unnecessary ones. As a result, better precision is achieved in less time. The threshold points values that are manually specified determine the various parameters required for MLP classifiers deployment.

A plant disease identification task by three deep learning meta-architectures such as Faster Region-based CNN(RCNN), Single Shot Multibox detector and Region based fully CNN(RFCN) (Muhammad et al.,2020). The suggested technique's primary goal is to find a confidence score showing the possibility that a right class is exists in a bounding box, in addition to detecting the existence of diseased and healthy ones. Transfer learning techniques are utilized to improve detection outcomes. All of the leaf classifications, whether healthy or damaged, were identified, demonstrating the work's uniqueness.

4. COMPARATIVE STUDY OF THE PAPERS REVIEWED

In this section, we have designed a table which provides the details about the methodology handled, advantages associated with each model, shortcomings faced with each of the approach and the metrices used for evaluation. Comparative table of research works is shown in Table 1.

Table 1. Comparative table of the research works

(Author, year)	Title	Methodology	Advantages	Issues	Metrics Used
Dhingra et al., (2019)	A novel Computer vision based neutrosophic approach for leaf disease identification and classification	**Segmentation:** Neutrosophic logic-based segmentation technique **Classification**: Random Forest classification	**Improved:** Segmentation and classification accuracy **Reduced:** Complexity and training time	Not capable of detecting specific diseases.	**Segmentation:** ROC curve Area under the curve (AUC) **Classification:** Accuracy Recall Precision Error rate Specificity
Vamsidhar et.al.,(2019)	Plant disease identification and classification using Image Processing	**Segmentation:** Hybrid K-means algorithm. **Classification:** SVM	**Improved:** Segmentation and classification accuracy **Reduced:** Loss landscape	Failed to detect the severity of the diseases	**Classification:** Accuracy
Nanehkaran et al. (2020)	Recognition of plant leaf diseases based on computer vision	**Segmentation:** HSI and LAB based hybrid algorithm **Classification:** SegCNN technique for classification.	**Improved:** Accuracy **Reduced:** Complexity, Random noise, complex background and uneven illumination	Structural changes are not reflected.	**Segmentation:** Run-time **Classification:** Accuracy & Recall Prediction probability
Louay Boukhris et al.(2020)	Tailored Deep Learning based Architecture for Smart Agriculture	**Detection** Mask-Region based convolutional neural network **Classification** Residual network architecture is used	**Improved:** Performance, Accuracy, Convergence speed. **Reduced:** Training time Error rate	Not applicable in real world applications.	**Classification** Accuracy **Detection** Precision Recall F1-Score

continued on following page

Table 1. Continued

(Author, year)	Title	Methodology	Advantages	Issues	Metrics Used
Hassan et al. (2020)	Identification of Plant-Leaf Diseases Using CNN and Transfer-Learning Approach	**Classification and Prediction**: Depth separable convolution architecture (EfficientNetB0)	**Improved:** Accuracy **Reduced:** Training time and Computational cost	Much sensitive to noise.	**Classification** Accuracy **Prediction** F1-score, Precision, recall
Huang et al., (2020)	Detection of Peach disease image based on Asymptotic Non-Local Means and PCNN-IPELM	**Denoising:** Asymptotic Non-Local Means (ANLM) image algorithm, **Feature extraction & Classification:** Parallel CNN, Extreme learning machine (ELM) optimized with the aid of linear particle swarm optimization (IPSO).	**Improved:** Denoising level, Accuracy & regression performance	Low convergence speed and classification accuracy.	**Classification** Accuracy
Yaonan Zhang et al., (2020)	Leaf Image Recognition Based on Bag of Features.	**Classification** Linear Support Vector Machine (SVM) **Feature Extraction** Bag Of Features (BOF), Dual-output pulse-coupled neural network (DPCNN), LDA (For texture features)	**Improved** Learning speed, computational cost. recognition accuracy. **Reduced** Learning cost and Iteration process reduced.	Low recognition accuracy.	**Classification** Accuracy
Sajad sabzi et al., (2020)	A Computer Vision System for the Automatic Classification of Five Varieties of Tree Leaf Images	**Classification** Hybrid ANN–ABC, hybrid ANN–BBO and Fisher linear discriminant analysis (LDA). **Feature Extraction** Hybrid ANN genetic algorithm, Hybrid ANN-DE and Gray Level Co-Occurrence matrix are used.	**Improved** accuracy **Reduced** Classifier error value, noise.	Not suitable for distinct classes feature extraction.	**Classification** Sensitivity, Specificity and accuracy, Receiver Operating characteristic (ROC) and AUC.
Narendra Narisetti et al., (2020)	Automated Spike Detection in Diverse European Wheat Plants Using Textural Features and the Frangi Filter in 2D Greenhouse Images.	**Classification** Wavelet based texture classification **Detection** Wheat spike detection using Laws Textural energy based neural network.	**Improved** accuracy and high throughput **Reduced** Error rate and ground truth data.	Performing timeseries analysis for a large population.	**Classification** Accuracy Sensitivity
Ramesh and Vydeki, (2020)	Recognition and classification of paddy leaf diseases using Optimized Deep Neural network with java algorithm	**Classification** Optimized deep neural network. **Segmentation** K-means clustering method.	**Improved** Prediction rate and stability.	Increased false classification rate.	**Segmentation** Accuracy and Precision **Classification** F1-Score TNR TPR FPR FNR FDR and NPV Prediction probability.
Aravind Krishnaswamy Rangarajan and Raja Purushothaman (2020)	Feature extraction is done by using pre-trained Visual Geometry Group 16(VGG16) architecture and classification is by using Multi Class Support Vector Machine (MSVM).	**Feature Extraction** Pre-trained VGG16 architecture **Classification** Multi Class Support Vector Machine (MSVM)	**Improved** Classification accuracy, Image enhancement, Productivity. **Reduced** Noise and classification time.	Limited availability of memory space.	**Classification** Accuracy

continued on following page

Table 1. Continued

(Author, year)	Title	Methodology	Advantages	Issues	Metrics Used
Samreen Naeem et al., (2021)	The Classification of Medicinal Plant Leaves Based on Multispectral and Texture Feature Using Machine Learning Approach.	**Classification** Multilayer Perceptron, LB, Bagging, RF and Simple Logistic **Edge detection** Sobel filter **Feature Selection** Chi-square based feature selection	**Improved** Accuracy. **Reduced** Noise and Time	Limited availability of image datasets.	**Classification** Accuracy, ROC curve, Recall and F-Measure.
Helong et al., (2021)	Optimized deep residual network system for diagnosing tomato pests.	**Disease Identification and Classification:** Image processing algorithm based on candidate hotspot detection and statistical inference methods	**Improved:** Optimized learning rate **Reduced:** Feature map scale and number of channels.	Difficult feature extraction capability.	**Detection:** Recall and Accuracy
Yibin et al.,(2022)	Rice diseases detection and classification using attention based neural network and Bayesian optimization.	**Detection and Classification:** Attention-based depth wise separable neural network with Bayesian optimization	**Improved:** Performance accuracy **Reduced:** Model size and platform management	Performance of the model using different optimization techniques and hyperparameter tuning needs to be checked. Convergence speed also needed to improve.	**Classification:** Accuracy
Elaraby et al., (2022)	Optimization of Deep Learning Model for Plant Disease Detection Using Particle Swarm Optimizer	**Optimization and Identification:** Alex Net Model **Classification:** **Alex Net Model**	**Improved:** Network structure and edge weights. **Reduced:** Loss and overfitting	Performance of the model is limited to particular set of datasets.	**Classification and Identification:** Sensitivity, Specificity, precision and F-1 Score
Hasan et al., (2023)	Design of Efficient Methods for the Detection of Tomato Leaf Disease Utilizing Proposed Ensemble CNN Model.	**Detection:** Ensemble CNN model	**Improved:** Performance **Reduced:** Overfitting and loss	A little noise in the sample images led to misclassification by the deep-learning model	**Detection:** Accuracy ROC Precision Recall F-1 Score

5.INFERENCES

This section presents the description of goals stated and achieved, Gaps associated with the study, various datasets used and the outcomes obtained with the methodology. Table 2 provides the inference of the study.

REFERENCES

Adams, J., Qiu, Y., Xu, Y., & Schnable, J. C. (2020). Plant segmentation by supervised machine learning methods. *The Plant Phenome Journal*, *3*(1), e20001. doi:10.1002/ppj2.20001

Table 2. Inference table

(Author, year)	Dataset	Goals stated and achieved	Reasons for achieving advantages	Gaps (reasons for shortfalls)
Dhingra et al. (2019)	Kapoor basil, Ram, Shyama basil Holy Basil plant image datasets	**Stated goals** Neutrosophic logic-based segmentation technique for disease identification. **Achieved goals** Greater accuracy.	Local structure of leaf well obtained in random forest algorithm.	Image quality needs to be increased.
Nanehkaran et al., (2020)	Plant village datasets.	**Stated goals** Disease symptom segmentation of Maize and Rice leaf images. **Achieved goals** Better segmentation effects and strong robustness.	1.Addition of max pooling layer. 2.Use of LAB threshold method.	Performance measures should be improved by adding some optimal classifiers.
Louay Boukhris et.al. (2020)	Plant Village datasets, COCO datasets	**Stated goals** Automatic damage detection, localization and identification. **Achieved goals** Able to classify the existence or absence of certain diseases.	Use of Adam optimizer and Transfer learning techniques.	More architecture complexity.
Muhammad Hammad Saleem et al., (2020)	MS COCO Datasets and Plant village datasets.	**Stated goals** Identification and localization of plant diseases. **Achieved goals** Better precision rates and faster computation time.	Adam Optimizer helped to improve average precision rates.	Mean average precision should be enhanced by adding better optimizers.
Yaonan Zhang et.al. (2020)	Flavia Datasets, ICL Datasets, Swedish leaf datasets, MEW datasets	**Stated goals** Bag of features and Dual Output pulse coupled neural network (BOF_DP) is used for classification. **Achieved goals** Improved training and classification speed achieved.	Use of Linear coding scheme (LLC).	Parameter optimization should be done to control the overall behavior of the architecture.
Huang et al., (2020)	Peach image datasets	**Stated goals** Peach disease detection by ANLM. **Achieved goals** Reduced interference of complex background in image	ANLM image denoising algorithm helped to remove noises.	Poor training effect due to similarity in some features of two diseases.
Samreen Naeem et.al (2020)	Medicinal plant leaves datasets.	**Stated goals** Select an efficient ML classifier and feature selection approach. **Achieved goals** Improved classification results.	Use of Chi-square feature selector and MLP classifier.	Instead of using pixel-based approach object-based approach can improve classification.
Sajad Sabzi et al., (2020)	Quince, river red gem, apple, mt. Atlas mastic tree, apricot datasets	**Stated goals** Automatic recognition and classification. **Achieved goals** Better accuracy and reduction in noise.	Use of Classifiers such as ANN_ABC, ANN_BBO and LDA.	LDA classifier is not suitable for large datasets.
Aravind Krishnaswamy and Raja Purushothaman, (2020)	Eggplant datasets.	**Stated goals** Create a standard dataset for eggplant and classify various diseases that can be affected. **Achieved goals** Better classification efficacy and accuracy.	VGG16 helped to achieve better classification efficiency.	Efficiency should be improved by adding more isolated leaf samples similar to eggplant.
Muhammad et al. (2021)	Plant village datasets	**Stated goals** Detection of tomato plant diseases. **Achieved goals** Better prediction accuracy.	Overfitting is avoided by using Modified U-net segmentation model.	Classification of some rare diseases or species is still problematic.
Helong et.al (2021)	IEEE CEC2017 datasets	**Stated goals** Optimized learning rate of the neural network **Achieved goals** Better learning rate is obtained using the proposed algorithm.	Improved fruit fly optimized algorithm helps to improve the searching of the learning rates.	Multiple hyperparameters has to be considered. Here only one hyperparameter evaluation is being considered.

continued on following page

Table 2. Continued

(Author, year)	Dataset	Goals stated and achieved	Reasons for achieving advantages	Gaps (reasons for shortfalls)
Yibin et.al (2022)	Rice leaf disease datasets.	**Stated goals** Rice disease classification and detection using attention based neural network and Bayesian optimization. **Achieved goals** Detection accuracy improved over mobile device environment also.	Attention-based depth wise separable neural network with Bayesian optimization (ADSNN-BO) is used for detection.	Performance of the model using different optimization techniques and hyperparameter tuning needs to be checked. Convergence speed also needed to improve.
Elarby et. al (2022)	Plantix dataset	**Stated Goals** Plant disease detection using particle swarm optimizer **Achieved goals** Greater detection and classification accuracy achieved.	Parameter control is being done with particle swarm optimization techniques.	Dataset size is being used is limited. Diverse dataset could be added.
Hasan et.al (2023)	Plant village dataset	**Stated goals** Detection of tomato leaf diseases using ensemble CNN models. **Achieved goals** Faster training and testing time and superior classification performance achieved.	Ensemble CNN model helped to achieve better detection accuracy.	A little noise in the sample images led to misclassification by the deep-learning model

Agarwal, M., Singh, A., Arjaria, S., Sinha, A., & Gupta, S. (2020, January 1). ToLeD: Tomato leaf disease detection using convolution neural network. *Procedia Computer Science*, *167*, 293–301. doi:10.1016/j.procs.2020.03.225

Al-Hiary, H., Bani-Ahmad, S., Reyalat, M., Braik, M., & Alrahamneh, Z. (2011). Fast and accurate detection and classification of plant diseases. *International Journal of Computer Applications*, *17*(1), 31–38. doi:10.5120/2183-2754

Bama, B. S., Valli, S. M., Raju, S., & Kumar, V. A. (2011, April). Content based leaf image retrieval (CBLIR) using shape, color and texture features. *Indian Journal of Computer Science and Engineering.*, *2*(2), 202–211.

Bock, C. H., Poole, G. H., Parker, P. E., & Gottwald, T. R. (2010, March 10). Plant disease severity estimated visually, by digital photography and image analysis, and by hyperspectral imaging. *Critical Reviews in Plant Sciences*, *29*(2), 59–107. doi:10.1080/07352681003617285

Boukhris L, Abderrazak JB, Besbes H. (2020). Tailored Deep Learning based Architecture for Smart Agriculture. In *2020 International Wireless Communications and Mobile Computing (IWCMC)*, (pp. 964-969). IEEE.

Camargo, A., & Smith, J. S. (2009, January 1). An image-processing based algorithm to automatically identify plant disease visual symptoms. *Biosystems Engineering*, *102*(1), 9–21. doi:10.1016/j.biosystemseng.2008.09.030

Camargo, A., & Smith, J. S. (2009, May 1). Image pattern classification for the identification of disease-causing agents in plants. *Computers and Electronics in Agriculture, 66*(2), 121–125. doi:10.1016/j.compag.2009.01.003

Chohan, M., Khan, A., Chohan, R., Katpar, S. H., & Mahar, M. S. (2020). Plant disease detection using deep learning. *International Journal of Recent Technology and Engineering., 9*(1), 909–914.

Cristin, R., Kumar, B. S., Priya, C., & Karthick, K. (2020, October). Deep neural network-based Rider-Cuckoo Search Algorithm for plant disease detection. *Artificial Intelligence Review, 53*(7), 4993–5018. doi:10.100710462-020-09813-w

DeCost, B. L., & Holm, E. A. (2015, December 1). A computer vision approach for automated analysis and classification of microstructural image data. *Computational Materials Science, 110,* 126–133. doi:10.1016/j.commatsci.2015.08.011

Dhingra, G., Kumar, V., & Joshi, H. D. (2019, March 1). A novel computer vision based neutrosophic approach for leaf disease identification and classification. *Measurement, 135,* 782–794. doi:10.1016/j.measurement.2018.12.027

Elaraby, A., Hamdy, W., & Alruwaili, M. (2022). Optimization of deep learning model for plant disease detection using particle swarm optimizer. *Computers, Materials & Continua, 71*(2), 4019–4031. doi:10.32604/cmc.2022.022161

Fu, P., Meacham-Hensold, K., Guan, K., & Bernacchi, C. J. (2019). Hyperspectral leaf reflectance as proxy for photosynthetic capacities: An ensemble approach based on multiple machine learning algorithms. *Frontiers in Plant Science, 10,* 730. doi:10.3389/fpls.2019.00730 PMID:31214235

Gocławski, J., Sekulska-Nalewajko, J., Gajewska, E., & Wielanek, M. (2009, December 1). An automatic segmentation method for scanned images of wheat root systems with dark discolourations. *International Journal of Applied Mathematics and Computer Science, 19*(4), 679–689. doi:10.2478/v10006-009-0055-x

Griffel, L. M., Delparte, D., & Edwards, J. (2018, October 1). Using Support Vector Machines classification to differentiate spectral signatures of potato plants infected with Potato Virus Y. *Computers and Electronics in Agriculture, 153,* 318–324. doi:10.1016/j.compag.2018.08.027

Guo, Y., Zhang, J., Yin, C., Hu, X., Zou, Y., Xue, Z., & Wang, W. (2020, August 18). Plant disease identification based on deep learning algorithm in smart farming. *Discrete Dynamics in Nature and Society, 2020,* 2020. doi:10.1155/2020/2479172

Han, L., Haleem, M. S., & Taylor, M. (2015). A novel computer vision-based approach to automatic detection and severity assessment of crop diseases. In *2015 Science and Information Conference (SAI)* (pp. 638-644). IEEE. 10.1109/SAI.2015.7237209

Hassan, S. M., Maji, A. K., Jasiński, M., Leonowicz, Z., & Jasińska, E. (2021, January). Identification of plant-leaf diseases using CNN and transfer-learning approach. *Electronics (Basel)*, *10*(12), 1388. doi:10.3390/electronics10121388

Hassan, S. M., Maji, A. K., Jasiński, M., Leonowicz, Z., & Jasińska, E. (2021, January). Identification of plant-leaf diseases using CNN and transfer-learning approach. *Electronics (Basel)*, *10*(12), 1388. doi:10.3390/electronics10121388

Huang, S., Zhou, G., He, M., Chen, A., Zhang, W., & Hu, Y. (2020, July 24). Detection of peach disease image based on asymptotic non-local means and PCNN-IPELM. *IEEE Access : Practical Innovations, Open Solutions*, *8*, 136421–136433. doi:10.1109/ACCESS.2020.3011685

Jamuna, K. S., Karpagavalli, S., Vijaya, M. S., Revathi, P., Gokilavani, S., & Madhiya, E. (2010). Classification of seed cotton yield based on the growth stages of cotton crop using machine learning techniques. In *International Conference on Advances in Computer Engineering,* (pp. 312-315). IEEE. 10.1109/ACE.2010.71

Korada, N.K., Kuma, N., & Deekshitulu, Y.V. (2012). Implementation of naïve Bayesian classifier and ada-boost algorithm using maize expert system. *International Journal of Information Sciences and techniques (IJIST), 2.*

Kothari, J. D. (2018). Plant Disease Identification using Artificial Intelligence: Machine Learning Approach. Jubin Dipakkumar Kothari (2018). Plant Disease Identification using Artificial Intelligence: Machine Learning Approach. *International Journal of Innovative Research in Computer and Communication Engineering.*, *7*(11), 11082–11085.

Krishnaswamy Rangarajan, A., & Purushothaman, R. (2020, February 11). Disease classification in eggplant using pre-trained VGG16 and MSVM. *Scientific Reports*, *10*(1), 1–1. doi:10.103841598-020-59108-x PMID:32047172

Lee, J. S., & Chung, Y. N. (2005, February 25). Integrating edge detection and thresholding approaches to segmenting femora and patellae from magnetic resonance images. *Biomedical Engineering: Applications, Basis and Communications.*, *17*(01), 1–1.

Mesquita, D. P. (2011). *Image analysis and chemometric techniques as monitoring tools to characterize aggregated and filamentous organisms in activated sludge processes.*

Mishra, P. K., & Bhusry, M. (2016). Artificial bee colony optimization based negative selection algorithms to classify iris plant dataset. *International Journal of Computer Applications*, *133*(10), 40–43. doi:10.5120/ijca2016912344

Mishra, S., Sachan, R., & Rajpal, D. (2020, January 1). Deep convolutional neural network-based detection system for real-time corn plant disease recognition. *Procedia Computer Science*, *167*, 2003–2010. doi:10.1016/j.procs.2020.03.236

Naeem, S., Ali, A., Chesneau, C., Tahir, M. H., Jamal, F., Sherwani, R. A., & Ul Hassan, M. (2021, February). The classification of medicinal plant leaves based on multispectral and texture feature using machine learning approach. *Agronomy (Basel)*, *11*(2), 263. doi:10.3390/agronomy11020263

Nanehkaran, Y. A., Zhang, D., Chen, J., Tian, Y., & Al-Nabhan, N. (2020, September 4). Recognition of plant leaf diseases based on computer vision. *Journal of Ambient Intelligence and Humanized Computing*, 1–8. doi:10.100712652-020-02505-x

Narisetti, N., Neumann, K., Röder, M. S., & Gladilin, E. (2020). Automated spike detection in diverse European wheat plants using textural features and the Frangi filter in 2D greenhouse images. [IEEE.]. *Frontiers in Plant Science*, *11*, 666. doi:10.3389/fpls.2020.00666 PMID:32655586

Pethybridge, S. J., & Nelson, S. C. (2015, October 29). Leaf Doctor: A new portable application for quantifying plant disease severity. *Plant Disease*, *99*(10), 1310–1316. doi:10.1094/PDIS-03-15-0319-RE PMID:30690990

Phadikar, S., Sil, J., & Das, A. K. (2013, January 1). Rice diseases classification using feature selection and rule generation techniques. *Computers and Electronics in Agriculture*, *90*, 76–85. doi:10.1016/j.compag.2012.11.001

Rahman, C. R., Arko, P. S., Ali, M. E., Khan, M. A., Apon, S. H., Nowrin, F., & Wasif, A. (2020, June 1). Identification and recognition of rice diseases and pests using convolutional neural networks. *Biosystems Engineering*, *194*, 112–120. doi:10.1016/j.biosystemseng.2020.03.020

Raji, A. O., & Alamutu, A. O. (2005, February 1). Prospects of Computer Vision Automated Sorting Systems in Agricultural Process Operations in Nigeria. *Agricultural Engineering International: CIGR Journal*.

Ramesh, S., & Vydeki, D. (2020). *Rice Disease Detection and Classification Using Deep Neural Network Algorithm. InMicro-Electronics and Telecommunication Engineering*. Springer.

Ramesh, S., & Vydeki, D. (2020, June 1). Recognition and classification of paddy leaf diseases using Optimized Deep Neural network with Jaya algorithm. *Information Processing in Agriculture, 7*(2), 249–260. doi:10.1016/j.inpa.2019.09.002

Rathore, N. P., & Prasad, L. (2020). Automatic rice plant disease recognition and identification using convolutional neural network. *Journal of Critical Reviews., 7*(15), 6076–6086.

Revathi, P., Revathi, R., & Hemalatha, M. (2011). Comparative study of knowledge in Crop diseases using Machine Learning Techniques. [IJCSIT]. *International Journal of Computer Science and Information Technologies, 2*(5), 2180–2182.

Sabzi, S., Pourdarbani, R., & Arribas, J. I. (2020, March). A computer vision system for the automatic classification of five varieties of tree leaf images. *Computers., 9*(1), 6. doi:10.3390/computers9010006

Saleem, M. H., Khanchi, S., Potgieter, J., & Arif, K. M. (2020, November). Image-based plant disease identification by deep learning meta-architectures. *Plants, 9*(11), 1451. doi:10.3390/plants9111451 PMID:33121188

Sena, D. G. Jr, Pinto, F. A., Queiroz, D. M., & Viana, P. A. (2003, August 1). Fall armyworm damaged maize plant identification using digital images. *Biosystems Engineering, 85*(4), 449–454. doi:10.1016/S1537-5110(03)00098-9

Thangaraj, R., Dinesh, D., Hariharan, S., Rajendar, S., Gokul, D., & Hariskarthi, T. R. (2020). Automatic recognition of avocado fruit diseases using modified deep convolutional neural network. *International Journal of Grid and Distributed Computing, 13*(1), 1550–1559.

Thanh Noi, P., & Kappas, M. (2017, December 22). Comparison of random forest, k-nearest neighbor, and support vector machine classifiers for land cover classification using Sentinel-2 imagery. *Sensors (Basel), 18*(1), 18. doi:10.339018010018 PMID:29271909

Tucker, C. C., & Chakraborty, S. (1997). Quantitative assessment of lesion characteristics and disease severity using digital image processing. *Journal of Phytopathology, 145*(7), 273–278. doi:10.1111/j.1439-0434.1997.tb00400.x

Ulutaş, H., & Aslantaş, V. (2023). Design of Efficient Methods for the Detection of Tomato Leaf Disease Utilizing Proposed Ensemble CNN Model. *Electronics (Basel), 12*(4), 827. doi:10.3390/electronics12040827

Vamsidhar, E., Rani, P. J., & Babu, K. R. (2019). Plant disease identification and classification using image processing. *International Journal of Engineering and Advanced Technology, 8*(3), 442–446.

Wang, Y., Wang, H., & Peng, Z. (2021). Rice diseases detection and classification using attention based neural network and bayesian optimization. *Expert Systems with Applications*, *178*, 114770. doi:10.1016/j.eswa.2021.114770

Plant IX. (n.dPlant IX. https://plantix.net/en/blog?page=1

The Mama Pirate. (2023). *Ten Plant Disease Identification Appps*. The Mama Pirate. https://themamapirate.com/plant-disease-identification-apps/

Yu, H., Liu, J., Chen, C., Heidari, A. A., Zhang, Q., & Chen, H. (2022). Optimized deep residual network system for diagnosing tomato pests. *Computers and Electronics in Agriculture*, *195*, 106805. doi:10.1016/j.compag.2022.106805

Zhang, Q., Pavlic, G., Chen, W., Fraser, R., Leblanc, S., & Cihlar, J. (2005). A semi-automatic segmentation procedure for feature extraction in remotely sensed imagery. *Computers & Geosciences*, *31*(3), 289–296. doi:10.1016/j.cageo.2004.10.003

Zhang, Y., Cui, J., Wang, Z., Kang, J., & Min, Y. (2020, January). Leaf image recognition based on bag of features. *Applied Sciences (Basel, Switzerland)*, *10*(15), 5177. doi:10.3390/app10155177

Chapter 10

Redefining Management With Advent of Artificial Intelligence in the Current Business World

Rohit Bhagat
University of Jammu, India

Vinay Chauhan
University of Jammu, India

ABSTRACT

The emergence of artificial intelligence has been a start of new era in the age of technology. The business world mostly relies on satisfying customer needs, with the introduction of artificial intelligence the task of marketer to satisfy customer has been made much simpler. The use of artificial intelligence has added a lot of value to the customers while purchasing, which has added to an overall increase in customer experience. Decision making is one of the most difficult tasks from the customer's point of view; when it comes to choosing a product, the use of artificial intelligence has been very fruitful in helping the customer in making decisions. The chapter tries to show how the use of artificial intelligence has changed the way of doing business. The chapter concludes with the future scope of artificial intelligence in the field of management and its application.

INTRODUCTION

The capability to perform difficult tasks in simpler way, a new technology has been invented having similarities to human intelligence, which is termed as artificial

DOI: 10.4018/978-1-6684-7659-8.ch010

intelligence. The ability of artificial intelligence to performs tasks with more accuracy makes this technology different from other technologies which were in use earlier. Artificial intelligence is termed as a computer-based technology which can use human like intelligence to interpret and solve problems. Artificial intelligence is a tool that uses machine learning language to make decisions with humans like intelligence and works with machines which are built to operate like a human being (Bilal & Oyedele, 2020; Monostori, 2003). Artificial intelligence has been described as a computer-based technology with scientific methodology for developing problem-based computer programmes (Ashrafian, 2017). Artificial intelligence is a machine learning language which is a computer-generated data mining programme specifically designed to predict the outcomes more accurately. Artificial intelligence comes with the advanced computer system proficiency which can sense, reason, and respond to their common real time problems with human-like intelligence (Abidoye & Chan, 2017).

Many earlier researches have shown that artificial intelligence have been on a rise from the past few years (Cioffi et al., 2020; Kuleto., 2021). The rise has been credited to not only in a particular field but across all the field in almost every domain. Many organisations have been employing artificial intelligence to understand customer needs in a more appropriate manner and solving them. In business world artificial intelligence has come as a boon in designing standard and reliable quality products which serves the consumer choices in a more efficient way (Arrieta et al., 2020). The artificial intelligence not only helps in producing a quality product but also at the same time maintaining low cost which helps in totally satisfying the customers (Mariani et al., 2022). The application of artificial intelligence in the business domain ranges from satisfying customers, demand prediction, inventory management, sales prediction, human resource management, increase in revenue, forecasting future decisions and many more. Artificial intelligence has been of huge importance in the field of consumer behaviour, specially understanding the parameters of psychological variables affecting the consumer decision making process (Rabby et al., 2021). Researchers have found that with help of artificial intelligence developed through machine learning language difficult decisions which require human like intelligence related to study and analysis of consumer behaviour could be achieved (Kim et al., 2023; Dias et al., 2023). The relation of marketing with artificial intelligence has widened the scope of marketer who can now use new and innovative ways to market products and gain benefits out of it.

Artificial intelligence has been described as a vital force for achieving overall development. Environmental social concern has grown at a rapid pace to an extent that the last decade was named as a "Decade of the Environment". In today's scenario, environmental concern has grown at a very rapid pace as far as citizens attitudinal and behavioural phenomena are concerned. The widespread and influence of the phenomenon can be seen in almost all human activities including business activities wherein one of the biggest challenges is to protect and preserve the earth's resources and the environment for the future generations without compromising the ability of the present generations to meet their own needs. As a result business have been realizing that their production and consumption has a direct impact on the environment therefore there is an urgent need to be concerned for the natural environment and subsequently initiate sustainable green practices involving consumers by enhancing their awareness for creating a favourable attitude and behavioural dispositions towards environment in which the human like intelligence technology of artificial intelligence comes as a blessing for the organisation to manufacture products according to the currents needs. Marketers have been using artificial intelligence technology to develop decision making capabilities related to every aspect of marketing activity (Nguyen, 2022). The current paper has tried to briefly introduce artificial intelligence having application in the business world. The paper further tries to study the trends and practices business organisations are using in the field of artificial intelligence towards enhancing marketing applicability. The study also puts forth the conceptual framework of artificial intelligence in relation to enhancing the efficiency of the business organisations.

Conceptual Framework of AI in Business World

Artificial intelligence in business world comes with a wider applicability. Artificial intelligence offers important potential benefits for consumers and their lifestyles. Business organisations have been using artificial intelligence to carry out main managerial activities which are planning, organising, leading and controlling. The application of artificial intelligence varies from formulating a strategy to how a business should be run to satisfying customer needs and wants (Coles, 1977; Galloway & Swiatek., 2018). The main advantage of artificial intelligence lies with delivering the decision with fairness in respect to the outcome of the decision (Jha & Topol, 2016). Artificial intelligence also follows procedural outcome fairness which means the fairness in terms of mechanism followed to arrive at the final decision. Artificial intelligence is a machine that behaves in a similar manner in which an intelligent human would have dealt the situation and came out with the solution (Lin et al., 2018). In the recent times artificial intelligence has significantly transformed fields such as biology, education, engineering, finance, and healthcare with its applicability and business organisations are no exception to it. The growing needs of consumers and lots of organisations competing to sell their product, the role of artificial intelligence becomes much more important as well as crucial. Artificial intelligence backed machine learning programmes and algorithms have demonstrated their effectiveness in processing large-scale and unstructured data in real-time, generating accurate predictions to assist decisions related to business world (Wirth, 2018; Soltaani & pooya, 2018). The applicability of artificial intelligence has turned out to be a milestone in terms of enhancing effectiveness in business performance (Dogru et al., 2020). The other main advantage of artificial intelligence enabled business activities has been its flexibility to adapt to situations and make decisions accordingly. Evidences reveal that increasing use of artificial intelligence has positively influenced the consumption pattern of products. The figure below depicts the pictorial representation showing benefits of artificial intelligence and its implication in field of business.

Consumer behaviour in true sense means the study of how individuals, groups, and organizations select, buy, use and dispose off goods, services, ideas or experiences to satisfy their needs and wants. Consumer behaviour is very complex, it is basically answering questions like what, when, why, where and how consumer makes a purchasing decision. Consumer goes for purchasing products having economies of scale in mind, some research reveals that consumer tend to go for purchase having more of satisfaction involved with the purchases, while some researches point out that consumers purchasing decision is more or less influenced by his family, friends and peer group (Panch et al., 2018). Consumer behaviour mainly revolved around the economic aspect of the purchase and value derived out of the purchase. Many researches have depicted that artificial intelligence enabled technology has come as an important tool in terms of understanding consumer behaviour and satisfying consumer needs (Lee & Park, 2022; Perihar & Yadav, 2021).

The introduction of artificial intelligence has been drastically transforming the business world. The activities related to business organisations vary from using

Figure 1. Benefits of artificial intelligence in business management

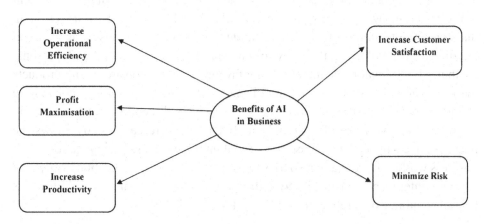

search alternatives for searching product till actually using the means to purchase the products and artificial intelligence has made all these activities much easier. The other way of looking at artificial intelligence enabled consumption means creating an environment which is suitable for living. It is also evident from the fact that consumers showing behaviour for artificial intelligence enabled technology for purchasing products play a very important role in contributing towards making environment sustainable for living (Omarov et al., 2022). As organisation heavily rely on demand among the consumers, the role of artificial intelligence becomes much more important. Organisations are now finding ways to produce products with help of artificial intelligence which are less costly and have value over a longer period of time (Teleaba, 2021). The past decade has shown a tremendous growth in use of artificial intelligence enabled technology in the business world. The researches show there are basically two ways of viewing it micro view and macro view. The micro view is basically concerned about the organisation's knowledge and awareness towards use of artificial intelligence in solving business related problems. The second view is macro view which basically deals with social, political, legal and technological factors having an effect on consumer's behaviour in relation to use of artificial intelligence enabled technology for decision making. Many organisations are coming forward and using artificial intelligence enabled technology in making products which are sustainable and profitable (Lehsen, 2020).

Consumers are well aware of artificial intelligence and are willing to pay extra from their pockets as they know that organisations using artificial intelligence enabled technology are the one selling products which are of high value and good quality (Limna, 2022; Furman & Seamans, 2019). The rapid growth of artificial intelligence in Indian economy has led foreign investors and global entrepreneurs to enter the market and make products valuable as well as according to current needs of customers (Wagner, 2020; Kopalle, 2022). Organisations using artificial intelligence enabled technology to produce products have to spend in the initial phase but the investment in initial phase pays way for significant results in the longer run. Researches claim that consumer tend to show positive attitude towards organisations using artificial intelligence technology in their day-to-day work. Values associated with personal attributes play a major role in developing favourable liking towards using products marketed through use of artificial intelligence technology. Consumer tends to have affinity towards organisations which are working in the direction towards improving their activities with use of artificial intelligence. The research study shows the positive affects by the emergence of artificial intelligence and its application to the business world. The research also tries to study the practical applications which the artificial intelligence has achieved and its impact on the business world towards overall economic development of the country.

Application of AI in Business World

The recent decade has seen a tremendous take off in the field of artificial intelligence. Organisations gaining early success with artificial intelligence are the organisations with managers having knowledge of artificial intelligence who can spot the problem and solve it effectively. Consumer attraction and profit maximisation are the biggest goals of the organisations, artificial intelligence has now made both the goals simpler and easier to achieve (Du et al., 2015). Artificial intelligence enabled technology help business organisations to access and forecast accurately the sales demand along with mechanism to satisfy customers. Organisations have been successfully using artificial intelligence to optimize the cost involved in sales which increases the profits of the organisations. Artificial intelligence has been useful in reducing the cost involved in serving and satisfying consumers. Artificial intelligence enabled technology helps in sorting the past data and predict future on the basis of analysis of past data which the organisation can use as per its needs (Fuster et al., 2022). The technology can also be used to manage its portfolio, manage risk, manage profit share and many more. The artificial intelligence also helps in managing the human resource of the organisation. It helps in screening the candidates suitable for job, improving process of recruitment, screening the applicants, conducting interviews through video conferencing and saving time and money in the process of recruitment and selection of employees. The major contribution of artificial intelligence is to transform the management practices in more effective and efficient manner (Luo et al., 2019).

The artificial intelligence also helps in managing data in both fronts be it data related to employee management or data related to customer satisfaction. Artificial intelligence is capable of handling past data related to sales, human resources, marketing, and customer demand and interprets and analyse the data having applicability in the current context. The results from the analysis help in profit maximization, sales maximization, resources optimization and effective utilisation of resources. Artificial intelligence algorithms help companies in inventory management. The machine generated algorithms helps in study and analysis of data related to stocks, their demand and supply relationship and accordingly predicted the inventory needed for the future course of action (Muller & Peres, 2019). This helps the managers in predicting future sales and maintain inventory accordingly. Artificial intelligence has come into very handy in terms of sorting of data and security associated with the data. The data studied and analysed through artificial intelligence is secure and safe for decision making. Many organisations are using artificial intelligence enabled technology in their financial activities as the technology removes chances of frauds and irregularities. As depicted in the figure below the study shows that their exists a procedural relationship between artificial intelligence enabled service and its impact on societal development.

Figure 2. Relevance of AI enabled services

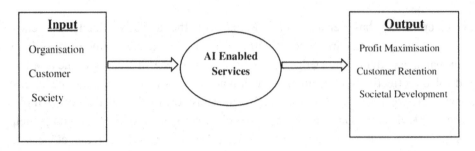

In today's time understanding customer preferences and satisfying customer is the most difficult job, when a lot options are available to the customer. The pace with which the marketing scenario is changing and competition is increasing rapidly on the global front, the importance of artificial intelligence increases many folds in the present context. The use of artificial intelligence has made the task easier for market researchers to accurately analyse what exactly the customer wants and how to satisfy the customer. The advent of artificial intelligence has made the job easier of the marketer to turn probable customer into end consumer (Ngai et al., 2009). The artificial intelligence also proves to be useful in managing customer data and selecting the loyal customer customers which helps the organisation in developing and better management of customer relationship management. Behaviour of a consumer is deeply impacted by the intention towards purchasing products and this purchase intention is deeply affected by way artificial intelligence moulds consumer behaviour favourably towards the product (Rabby et al., 2021). Consumers purchasing products always look for companies which show concern towards their problems and are willing to provide them with valuable offerings, this problem of the manufacturer has been well taken care by the artificial intelligence enabled technology to solve customer problems (Khan et al., 2022). Consumers always show favourable intention towards organisations offering solutions to their purchase related problems through artificial intelligence enabled technology.

Another important field in which artificial intelligence has revolutionised tremendously is online retailing. The use of artificial intelligence has made online retailing much more lucrative and profit earning for the firms doing online businesses (Weber & Schutte, 2019). The use of artificial intelligence has made e-retailing firms to connect in a better manner with the consumers and instantly solve their queries through chatbots and sell the desired product to the customers. The interactions between chatbots and consumers solves consumer problems in no time with human like intelligence capability of decision making. The artificial intelligence enabled technology has not only made buying and selling process simpler but also saved time and money of both customers as well as online retailing firms. The use of artificial intelligence technology in online retailing has enhanced the profits of firms tremendously with many online organisations going for using artificial intelligence technology in their daily routine. The use of artificial intelligence in the field of marketing is regarded as one of the most important applications of artificial intelligence. Artificial intelligence in marketing includes techniques that can contribute tremendously in predicting needs and wants of new customers and accurately personalizing and recommending better and suitable options having value to the customers.

CONCLUSION

The introduction of Artificial intelligence technology has come as a lifeline in reshaping business and economy by satisfying consumers needs and maximising profits of the firms. The major contribution of artificial intelligence has been satisfying consumer needs and wants, which is only possible through technological advancements. The pace with which artificial intelligence is progressing the day is not far when most of activities will be performed with artificial intelligence enabled technology. The knowledge of artificial intelligence will benefit marketers as well as researchers through its applicability in different domains and fields. Artificial intelligence has been a pathbreaking technology in the fields of online retailing and financial industries towards improving customer satisfaction and profit maximisation. Artificial intelligence capable of proactively taking actions for assisting consumers in every activity will become the norms in near future, thus widening the scope for formulating new ways and mechanism of satisfying consumers. Artificial intelligence is becoming more intelligent with every passing day and it is expected that in future the consumers would find their lifestyle much different and sustainable. The current research study on Artificial intelligence has put forth that the technology will act as a strategic differentiator among competing business organisations, and also will act as a greater strength in developing economies of scale. A society based on superintelligent artificial intelligence along with its benefits brings lots of responsibility for the stakeholders to use the technology for benefit of socio-economic development of the nation.

REFERENCES

Abidoye, R. B., & Chan, A. P. (2017). Valuers' receptiveness to the application of artificial intelligence in property valuation. *Pacific Rim Property Research Journal*, *23*(2), 175–193. doi:10.1080/14445921.2017.1299453

Arrieta, A. B., Díaz-Rodríguez, N., Del Ser, J., Bennetot, A., Tabik, S., Barbado, A., & Herrera, F. (2020). Explainable Artificial Intelligence (XAI): Concepts, taxonomies, opportunities and challenges toward responsible AI. *Information Fusion*, *58*, 82–115. doi:10.1016/j.inffus.2019.12.012

Ashrafian, H. (2017). Can artificial intelligences suffer from mental illness? A philosophical matter to consider. *Science and Engineering Ethics*, *23*(2), 403–412. doi:10.100711948-016-9783-0 PMID:27351772

Bilal, M., & Oyedele, L. O. (2020). Big Data with deep learning for benchmarking profitability performance in project tendering. *Expert Systems with Applications*, *147*, 113194. doi:10.1016/j.eswa.2020.113194

Coles, L. S. (1977). The application of artificial intelligence to medicine. *Futures*, *9*(4), 315–323. doi:10.1016/0016-3287(77)90097-0

Dias, T., Gonçalves, R., da Costa, R. L., Pereira, L. F., & Dias, Á. (2023). The impact of artificial intelligence on consumer behaviour and changes in business activity due to pandemic effects. *Human Technology*, *19*(1), 121–148. doi:10.14254/1795-6889.2023.19-1.8

Dogru, A. K., & Keskin, B. B. (2020). AI in operations management: Applications, challenges and opportunities. *Journal of Data. Information & Management*, *2*, 67–74.

Du, R. Y., Hu, Y., & Damangir, S. (2015). Leveraging trends in online searches for product features in market response modeling. *Journal of Marketing*, *79*(1), 29–43. doi:10.1509/jm.12.0459

Furman, J., & Seamans, R. (2019). AI and the Economy. *Innovation Policy and the Economy*, *19*(1), 161–191. doi:10.1086/699936

Fuster, A., Goldsmith-Pinkham, P., Ramadorai, T., & Walther, A. (2022). Predictably unequal? The effects of machine learning on credit markets. *The Journal of Finance*, *77*(1), 5–47. doi:10.1111/jofi.13090

Galloway, C., & Swiatek, L. (2018). Public relations and artificial intelligence: It's not (just) about robots. *Public Relations Review*, *44*(5), 734–740. doi:10.1016/j.pubrev.2018.10.008

Jha, S., & Topol, E. J. (2016). Adapting to artificial intelligence: Radiologists and pathologists as information specialists. *Journal of the American Medical Association*, *316*(22), 2353–2354. doi:10.1001/jama.2016.17438 PMID:27898975

Khan, S., Tomar, S., Fatima, M., & Khan, M. Z. (2022). Impact of artificial intelligent and industry 4.0 based products on consumer behaviour characteristics: A meta-analysis-based review. *Sustainable Operations and Computers*, *3*, 218–225. doi:10.1016/j.susoc.2022.01.009

Kim, T., Lee, H., Kim, M. Y., Kim, S., & Duhachek, A. (2023). AI increases unethical consumer behavior due to reduced anticipatory guilt. *Journal of the Academy of Marketing Science*, *51*(4), 785–801. doi:10.100711747-021-00832-9

Kopalle, P. K., Gangwar, M., Kaplan, A., Ramachandran, D., Reinartz, W., & Rindfleisch, A. (2022). Examining artificial intelligence (AI) technologies in marketing via a global lens: Current trends and future research opportunities. *International Journal of Research in Marketing*, *39*(2), 522–540. doi:10.1016/j.ijresmar.2021.11.002

Kuleto, V., Ilić, M., Dumangiu, M., Ranković, M., Martins, O. M., Păun, D., & Mihoreanu, L. (2021). Exploring opportunities and challenges of artificial intelligence and machine learning in higher education institutions. *Sustainability (Basel)*, *13*(18), 10424. doi:10.3390u131810424

Lahsen, M. (2020). Should AI be designed to save us from ourselves?: Artificial intelligence for sustainability. *IEEE Technology and Society Magazine*, *39*(2), 60–67. doi:10.1109/MTS.2020.2991502

Lee, M., & Park, J. S. (2022). Do parasocial relationships and the quality of communication with AI shopping chatbots determine middle-aged women consumers' continuance usage intentions? *Journal of Consumer Behaviour*, *21*(4), 842–854. doi:10.1002/cb.2043

Limna, P. (2022). Artificial Intelligence (AI) in the hospitality industry: A review article. *Int. J. Comput. Sci. Res*, *6*, 1–12.

Lin, P. H., Wooders, A., Wang, J. T. Y., & Yuan, W. M. (2018). Artificial intelligence, the missing piece of online education? *IEEE Engineering Management Review*, *46*(3), 25–28. doi:10.1109/EMR.2018.2868068

Luo, X., Tong, S., Fang, Z., & Qu, Z. (2019). Frontiers: Machines vs. humans: The impact of artificial intelligence chatbot disclosure on customer purchases. *Marketing Science*, *38*(6), 937–947. doi:10.1287/mksc.2019.1192

Mariani, M. M., Perez-Vega, R., & Wirtz, J. (2022). AI in marketing, consumer research and psychology: A systematic literature review and research agenda. *Psychology and Marketing*, *39*(4), 755–776. doi:10.1002/mar.21619

Monostori, L. (2003). AI and machine learning techniques for managing complexity, changes and uncertainties in manufacturing. *Engineering Applications of Artificial Intelligence*, *16*(4), 277–291. doi:10.1016/S0952-1976(03)00078-2

Muller, E., & Peres, R. (2019). The effect of social networks structure on innovation performance: A review and directions for research. *International Journal of Research in Marketing*, *36*(1), 3–19. doi:10.1016/j.ijresmar.2018.05.003

Ngai, E. W., Xiu, L., & Chau, D. C. (2009). Application of data mining techniques in customer relationship management: A literature review and classification. *Expert Systems with Applications*, *36*(2), 2592–2602. doi:10.1016/j.eswa.2008.02.021

Nguyen, T. M., Quach, S., & Thaichon, P. (2022). The effect of AI quality on customer experience and brand relationship. *Journal of Consumer Behaviour*, *21*(3), 481–493. doi:10.1002/cb.1974

Omarov, B., Tursynbayev, A., Zhakypbekova, G., & Beissenova, G. (2022). Effect Of Chatbots In Digital Marketing To Perceive The Consumer Behaviour. *Journal of Positive School Psychology*, *6*(8), 7143–7161.

Panch, T., Szolovits, P., & Atun, R. (2018). Artificial intelligence, machine learning and health systems. *Journal of Global Health*, *8*(2), 020303. doi:10.7189/jogh.08.020303 PMID:30405904

Parihar, V., & Yadav, S. (2021). Comparison estimation of effective consumer future preferences with the application of AI. *Vivekananda Journal of Research*, *10*, 133–145.

Rabby, F., Chimhundu, R., & Hassan, R. (2021). Artificial intelligence in digital marketing influences consumer behaviour: A review and theoretical foundation for future research. *Academy of Marketing Studies Journal*, *25*(5), 1–7.

Soltani-Fesaghandis, G., & Pooya, A. (2018). Design of an artificial intelligence system for predicting success of new product development and selecting proper market-product strategy in the food industry. *The International Food and Agribusiness Management Review*, *21*(7), 847–864. doi:10.22434/IFAMR2017.0033

Teleaba, F., Popescu, S., Olaru, M., & Pitic, D. (2021). Risks of observable and unobservable biases in artificial intelligence used for predicting consumer choice. *Amfiteatru Economic*, *23*(56), 102–119. doi:10.24818/EA/2021/56/102

Wagner, D. N. (2020). Economic patterns in a world with artificial intelligence. *Evolutionary and Institutional Economics Review*, *17*(1), 111–131. doi:10.100740844-019-00157-x

Weber, F. D., & Schutte, R. (2019). State-of-the-art and adoption of artificial intelligence in retailing. *Digital Policy*. *Regulation & Governance*, *21*(3), 264–279. doi:10.1108/DPRG-09-2018-0050

Wirth, N. (2018). Hello marketing, what can artificial intelligence help you with? *International Journal of Market Research*, *60*(5), 435–438. doi:10.1177/1470785318776841

Compilation of References

Bama, B. S., Valli, S. M., Raju, S., & Kumar, V. A. (2011, April). Content based leaf image retrieval (CBLIR) using shape, color and texture features. *Indian Journal of Computer Science and Engineering.*, 2(2), 202–211.

Gocławski, J., Sekulska-Nalewajko, J., Gajewska, E., & Wielanek, M. (2009, December 1). An automatic segmentation method for scanned images of wheat root systems with dark discolourations. *International Journal of Applied Mathematics and Computer Science*, 19(4), 679–689. doi:10.2478/v10006-009-0055-x

Camargo, A., & Smith, J. S. (2009, May 1). Image pattern classification for the identification of disease-causing agents in plants. *Computers and Electronics in Agriculture*, 66(2), 121–125. doi:10.1016/j.compag.2009.01.003

Jamuna, K. S., Karpagavalli, S., Vijaya, M. S., Revathi, P., Gokilavani, S., & Madhiya, E. (2010). Classification of seed cotton yield based on the growth stages of cotton crop using machine learning techniques. In *International Conference on Advances in Computer Engineering,* (pp. 312-315). IEEE. 10.1109/ACE.2010.71

Korada, N.K., Kuma, N., & Deekshitulu, Y.V. (2012). Implementation of naïve Bayesian classifier and ada-boost algorithm using maize expert system. *International Journal of Information Sciences and techniques (IJIST)*, 2.

Revathi, P., Revathi, R., & Hemalatha, M. (2011). Comparative study of knowledge in Crop diseases using Machine Learning Techniques. [IJCSIT]. *International Journal of Computer Science and Information Technologies*, 2(5), 2180–2182.

Phadikar, S., Sil, J., & Das, A. K. (2013, January 1). Rice diseases classification using feature selection and rule generation techniques. *Computers and Electronics in Agriculture*, 90, 76–85. doi:10.1016/j.compag.2012.11.001

Pethybridge, S. J., & Nelson, S. C. (2015, October 29). Leaf Doctor: A new portable application for quantifying plant disease severity. *Plant Disease*, 99(10), 1310–1316. doi:10.1094/PDIS-03-15-0319-RE PMID:30690990

DeCost, B. L., & Holm, E. A. (2015, December 1). A computer vision approach for automated analysis and classification of microstructural image data. *Computational Materials Science, 110,* 126–133. doi:10.1016/j.commatsci.2015.08.011

Han, L., Haleem, M. S., & Taylor, M. (2015). A novel computer vision-based approach to automatic detection and severity assessment of crop diseases. In *2015 Science and Information Conference (SAI)* (pp. 638-644). IEEE. 10.1109/SAI.2015.7237209

Tucker, C. C., & Chakraborty, S. (1997). Quantitative assessment of lesion characteristics and disease severity using digital image processing. *Journal of Phytopathology, 145*(7), 273–278. doi:10.1111/j.1439-0434.1997.tb00400.x

Mishra, P. K., & Bhusry, M. (2016). Artificial bee colony optimization based negative selection algorithms to classify iris plant dataset. *International Journal of Computer Applications, 133*(10), 40–43. doi:10.5120/ijca2016912344

Griffel, L. M., Delparte, D., & Edwards, J. (2018, October 1). Using Support Vector Machines classification to differentiate spectral signatures of potato plants infected with Potato Virus Y. *Computers and Electronics in Agriculture, 153,* 318–324. doi:10.1016/j.compag.2018.08.027

Kothari, J. D. (2018). Plant Disease Identification using Artificial Intelligence: Machine Learning Approach. Jubin Dipakkumar Kothari (2018). Plant Disease Identification using Artificial Intelligence: Machine Learning Approach. *International Journal of Innovative Research in Computer and Communication Engineering., 7*(11), 11082–11085.

Thanh Noi, P., & Kappas, M. (2017, December 22). Comparison of random forest, k-nearest neighbor, and support vector machine classifiers for land cover classification using Sentinel-2 imagery. *Sensors (Basel), 18*(1), 18. doi:10.339018010018 PMID:29271909

Fu, P., Meacham-Hensold, K., Guan, K., & Bernacchi, C. J. (2019). Hyperspectral leaf reflectance as proxy for photosynthetic capacities: An ensemble approach based on multiple machine learning algorithms. *Frontiers in Plant Science, 10,* 730. doi:10.3389/fpls.2019.00730 PMID:31214235

Plant IX. (n.d Plant IX. https://plantix.net/en/blog?page=1

The Mama Pirate. (2023). *Ten Plant Disease Identification Appps.* The Mama Pirate. https://themamapirate.com/plant-disease-identification-apps/

Guo, Y., Zhang, J., Yin, C., Hu, X., Zou, Y., Xue, Z., & Wang, W. (2020, August 18). Plant disease identification based on deep learning algorithm in smart farming. *Discrete Dynamics in Nature and Society, 2020,* 2020. doi:10.1155/2020/2479172

Chohan, M., Khan, A., Chohan, R., Katpar, S. H., & Mahar, M. S. (2020). Plant disease detection using deep learning. *International Journal of Recent Technology and Engineering., 9*(1), 909–914.

Cristin, R., Kumar, B. S., Priya, C., & Karthick, K. (2020, October). Deep neural network-based Rider-Cuckoo Search Algorithm for plant disease detection. *Artificial Intelligence Review, 53*(7), 4993–5018. doi:10.100710462-020-09813-w

Bock, C. H., Poole, G. H., Parker, P. E., & Gottwald, T. R. (2010, March 10). Plant disease severity estimated visually, by digital photography and image analysis, and by hyperspectral imaging. *Critical Reviews in Plant Sciences*, *29*(2), 59–107. doi:10.1080/07352681003617285

Mishra, S., Sachan, R., & Rajpal, D. (2020, January 1). Deep convolutional neural network-based detection system for real-time corn plant disease recognition. *Procedia Computer Science*, *167*, 2003–2010. doi:10.1016/j.procs.2020.03.236

Ramesh, S., & Vydeki, D. (2020). *Rice Disease Detection and Classification Using Deep Neural Network Algorithm. InMicro-Electronics and Telecommunication Engineering.* Springer.

Hassan, S. M., Maji, A. K., Jasiński, M., Leonowicz, Z., & Jasińska, E. (2021, January). Identification of plant-leaf diseases using CNN and transfer-learning approach. *Electronics (Basel)*, *10*(12), 1388. doi:10.3390/electronics10121388

Rathore, N. P., & Prasad, L. (2020). Automatic rice plant disease recognition and identification using convolutional neural network. *Journal of Critical Reviews.*, *7*(15), 6076–6086.

Rahman, C. R., Arko, P. S., Ali, M. E., Khan, M. A., Apon, S. H., Nowrin, F., & Wasif, A. (2020, June 1). Identification and recognition of rice diseases and pests using convolutional neural networks. *Biosystems Engineering*, *194*, 112–120. doi:10.1016/j.biosystemseng.2020.03.020

Agarwal, M., Singh, A., Arjaria, S., Sinha, A., & Gupta, S. (2020, January 1). ToLeD: Tomato leaf disease detection using convolution neural network. *Procedia Computer Science*, *167*, 293–301. doi:10.1016/j.procs.2020.03.225

Adams, J., Qiu, Y., Xu, Y., & Schnable, J. C. (2020). Plant segmentation by supervised machine learning methods. *The Plant Phenome Journal*, *3*(1), e20001. doi:10.1002/ppj2.20001

Dhingra, G., Kumar, V., & Joshi, H. D. (2019, March 1). A novel computer vision based neutrosophic approach for leaf disease identification and classification. *Measurement*, *135*, 782–794. doi:10.1016/j.measurement.2018.12.027

Nanehkaran, Y. A., Zhang, D., Chen, J., Tian, Y., & Al-Nabhan, N. (2020, September 4). Recognition of plant leaf diseases based on computer vision. *Journal of Ambient Intelligence and Humanized Computing*, 1–8. doi:10.100712652-020-02505-x

Boukhris L, Abderrazak JB, Besbes H. (2020). Tailored Deep Learning based Architecture for Smart Agriculture. In *2020 International Wireless Communications and Mobile Computing (IWCMC)*, (pp. 964-969). IEEE.

Lee, J. S., & Chung, Y. N. (2005, February 25). Integrating edge detection and thresholding approaches to segmenting femora and patellae from magnetic resonance images. *Biomedical Engineering: Applications, Basis and Communications.*, *17*(01), 1–1.

Huang, S., Zhou, G., He, M., Chen, A., Zhang, W., & Hu, Y. (2020, July 24). Detection of peach disease image based on asymptotic non-local means and PCNN-IPELM. *IEEE Access : Practical Innovations, Open Solutions*, *8*, 136421–136433. doi:10.1109/ACCESS.2020.3011685

Zhang, Y., Cui, J., Wang, Z., Kang, J., & Min, Y. (2020, January). Leaf image recognition based on bag of features. *Applied Sciences (Basel, Switzerland)*, *10*(15), 5177. doi:10.3390/app10155177

Sabzi, S., Pourdarbani, R., & Arribas, J. I. (2020, March). A computer vision system for the automatic classification of five varieties of tree leaf images. *Computers.*, *9*(1), 6. doi:10.3390/computers9010006

Narisetti, N., Neumann, K., Röder, M. S., & Gladilin, E. (2020). Automated spike detection in diverse European wheat plants using textural features and the Frangi filter in 2D greenhouse images. [IEEE.]. *Frontiers in Plant Science*, *11*, 666. doi:10.3389/fpls.2020.00666 PMID:32655586

Ramesh, S., & Vydeki, D. (2020, June 1). Recognition and classification of paddy leaf diseases using Optimized Deep Neural network with Jaya algorithm. *Information Processing in Agriculture*, *7*(2), 249–260. doi:10.1016/j.inpa.2019.09.002

Krishnaswamy Rangarajan, A., & Purushothaman, R. (2020, February 11). Disease classification in eggplant using pre-trained VGG16 and MSVM. *Scientific Reports*, *10*(1), 1–1. doi:10.103841598-020-59108-x PMID:32047172

Naeem, S., Ali, A., Chesneau, C., Tahir, M. H., Jamal, F., Sherwani, R. A., & Ul Hassan, M. (2021, February). The classification of medicinal plant leaves based on multispectral and texture feature using machine learning approach. *Agronomy (Basel)*, *11*(2), 263. doi:10.3390/agronomy11020263

Saleem, M. H., Khanchi, S., Potgieter, J., & Arif, K. M. (2020, November). Image-based plant disease identification by deep learning meta-architectures. *Plants*, *9*(11), 1451. doi:10.3390/plants9111451 PMID:33121188

Vamsidhar, E., Rani, P. J., & Babu, K. R. (2019). Plant disease identification and classification using image processing. *International Journal of Engineering and Advanced Technology*, *8*(3), 442–446.

Mesquita, D. P. (2011). *Image analysis and chemometric techniques as monitoring tools to characterize aggregated and filamentous organisms in activated sludge processes.*

Thangaraj, R., Dinesh, D., Hariharan, S., Rajendar, S., Gokul, D., & Hariskarthi, T. R. (2020). Automatic recognition of avocado fruit diseases using modified deep convolutional neural network. *International Journal of Grid and Distributed Computing*, *13*(1), 1550–1559.

Yu, H., Liu, J., Chen, C., Heidari, A. A., Zhang, Q., & Chen, H. (2022). Optimized deep residual network system for diagnosing tomato pests. *Computers and Electronics in Agriculture*, *195*, 106805. doi:10.1016/j.compag.2022.106805

Wang, Y., Wang, H., & Peng, Z. (2021). Rice diseases detection and classification using attention based neural network and bayesian optimization. *Expert Systems with Applications*, *178*, 114770. doi:10.1016/j.eswa.2021.114770

Elaraby, A., Hamdy, W., & Alruwaili, M. (2022). Optimization of deep learning model for plant disease detection using particle swarm optimizer. *Computers, Materials & Continua*, *71*(2), 4019–4031. doi:10.32604/cmc.2022.022161

Ulutaş, H., & Aslantaş, V. (2023). Design of Efficient Methods for the Detection of Tomato Leaf Disease Utilizing Proposed Ensemble CNN Model. *Electronics (Basel)*, *12*(4), 827. doi:10.3390/electronics12040827

Al-Hiary, H., Bani-Ahmad, S., Reyalat, M., Braik, M., & Alrahamneh, Z. (2011). Fast and accurate detection and classification of plant diseases. *International Journal of Computer Applications*, *17*(1), 31–38. doi:10.5120/2183-2754

Sena, D. G. Jr, Pinto, F. A., Queiroz, D. M., & Viana, P. A. (2003, August 1). Fall armyworm damaged maize plant identification using digital images. *Biosystems Engineering*, *85*(4), 449–454. doi:10.1016/S1537-5110(03)00098-9

Raji, A. O., & Alamutu, A. O. (2005, February 1). Prospects of Computer Vision Automated Sorting Systems in Agricultural Process Operations in Nigeria. *Agricultural Engineering International: CIGR Journal*.

Zhang, Q., Pavlic, G., Chen, W., Fraser, R., Leblanc, S., & Cihlar, J. (2005). A semi-automatic segmentation procedure for feature extraction in remotely sensed imagery. *Computers & Geosciences*, *31*(3), 289–296. doi:10.1016/j.cageo.2004.10.003

Camargo, A., & Smith, J. S. (2009, January 1). An image-processing based algorithm to automatically identify plant disease visual symptoms. *Biosystems Engineering*, *102*(1), 9–21. doi:10.1016/j.biosystemseng.2008.09.030

Abidoye, R. B., & Chan, A. P. (2017). Valuers' receptiveness to the application of artificial intelligence in property valuation. *Pacific Rim Property Research Journal*, *23*(2), 175–193. doi:10.1080/14445921.2017.1299453

Acharjee, S. (2016). A semiautomated approach using GUI for the detection of red blood cells. *2016 International conference on electrical, electronics, and optimization techniques (ICEEOT)*. IEEE. 10.1109/ICEEOT.2016.7755669

Adegun, A. A., & Viriri, S. (2020). FCN-based DenseNet framework for automated detection and classification of skin lesions in dermoscopy images. *IEEE Access: Practical Innovations, Open Solutions*, *8*, 150377–150396. doi:10.1109/ACCESS.2020.3016651

Akbar, S., Hassan, S. A., Shoukat, A., Alyami, J., & Bahaj, S. A. (2022). Detection of microscopic glaucoma through fundus images using deep transfer learning approach. *Microscopy Research and Technique*, *85*(6), 2259–2276. doi:10.1002/jemt.24083 PMID:35170136

Akkus, Z., Galimzianova, A., Hoogi, A., Rubin, D. L., & Erickson, B. J. (2017). Deep learning for brain mri segmentation: State of the art and future directions. *Journal of Digital Imaging*, *30*(4), 1–11. doi:10.100710278-017-9983-4 PMID:28577131

Alberts, B. (2003). Molecular biology of the cell. *Scandinavian Journal of Rheumatology*, *32*(2), 125–125.

Alice Nithya, C. Lakshmi. (2015). Iris recognition techniques: A Literature Survey, Project: Multi-Unit Feature Level Fusion Approach Using PPCA, July 2015. *International Journal of Applied Engineering Research: IJAER, 10*(12), 32525–32546.

Alomari, Y. M., Sheikh Abdullah, S. N. H., Zaharatul Azma, R., & Omar, K. (2014). Automatic detection and quantification of WBCs and RBCs using iterative structured circle detection algorithm. *Computational and Mathematical Methods in Medicine, 2014*, 2014. doi:10.1155/2014/979302 PMID:24803955

AlZahrani, F., Crosby, M., & Fiorillo, L. (2019, March 1). Use of AccuVein AV400 for identification of probable RICH. *JAAD Case Reports, 5*(3), 213–215. doi:10.1016/j.jdcr.2018.11.022 PMID:30809562

American Cancer Society. (n.d.). *Cancer facts & figures 2021*. American Cancer Society. https://www.cancer.org/research/cancer-facts-statistics/all-cancer-facts-figures/cancer-facts-figures-2021.html

Anoj, S., Tri Dev, A., & Ha, L. (2018). Evaluation of Water Indices for Surface Water Extraction in a Landsat 8 Scene of Nepal. *Sensors, 18*, 1-15.

Arinaldi, A., Pradana, J., & Gurusinga, A. (2018). Detection and classification of vehicles for traffic video analytics. *Procedia Computer Science, 144*, 259–268. doi:10.1016/j.procs.2018.10.527

Arrieta, A. B., Díaz-Rodríguez, N., Del Ser, J., Bennetot, A., Tabik, S., Barbado, A., & Herrera, F. (2020). Explainable Artificial Intelligence (XAI): Concepts, taxonomies, opportunities and challenges toward responsible AI. *Information Fusion, 58*, 82–115. doi:10.1016/j.inffus.2019.12.012

Arslan, Ö. & Karhan, M. (2022). Effect of Hilbert-Huang transform on classification of PCG signals using machine learning. *Journal of King Saud University-Computer and Information Sciences.*

Ashrafian, H. (2017). Can artificial intelligences suffer from mental illness? A philosophical matter to consider. *Science and Engineering Ethics, 23*(2), 403–412. doi:10.100711948-016-9783-0 PMID:27351772

Aulagnier, J., Hoc, C., Mathieu, E., Dreyfus, J. F., Fischler, M., & Le Guen, M. (2014, August). Efficacy of AccuVein to facilitate peripheral intravenous placement in adults presenting to an emergency department: A randomized clinical trial. *Academic Emergency Medicine, 21*(8), 858–863. doi:10.1111/acem.12437 PMID:25176152

Aznag, K., Datsi, T., El oirrak, A., & El bachari, E. (2020). Ahmed El oirrak & Essaid El bachari (2020), Binary image description using frequent itemsets. *Journal of Big Data, 7*(1), 32. doi:10.118640537-020-00307-8

Badrinarayanan, V., Kendall, A., & Cipolla, R. (2017). SegNet: A deep convolutional encoder-decoder architecture for image segmentation. *IEEE Transactions on Pattern Analysis and Machine Intelligence, 39*(12), 2481–2495. doi:10.1109/TPAMI.2016.2644615 PMID:28060704

Baker, S., & Kanade, T. (2000). Hallucinating faces. *Proc. Fourth IEEE Int. Conf. on Automatic Face and Gesture Recognition.* 10.1109/AFGR.2000.840616

Banerjee, S., & Mitra, S. (2020). Novel Volumetric Sub-region Segmentation in Brain Tumors. *Frontiers in Computational Neuroscience, 14*, 3. doi:10.3389/fncom.2020.00003 PMID:32038216

Ben Aziza, S., Dzahini, D., & Gallin-Martel, L. (2015). A high speed high resolution readout with 14-bits area efficient SAR-ADC adapted for new generations of CMOS image sensors. *2015 11th Conference on Ph.D. Research in Microelectronics and Electronics (PRIME),* (pp. 89-92). 10.1109/PRIME.2015.7251341

Bian, W., Chen, Y., Ye, X., & Zhang, Q. (2021). An Optimization-Based Meta-Learning Model for MRI Reconstruction with Diverse Dataset. *Journal of Imaging, 7*(11), 231. doi:10.3390/jimaging7110231 PMID:34821862

Bilal, M., & Oyedele, L. O. (2020). Big Data with deep learning for benchmarking profitability performance in project tendering. *Expert Systems with Applications, 147*, 113194. doi:10.1016/j.eswa.2020.113194

Blum T, Kleeberger V, Bichlmeier C, & Navab N. (2012). Mirracle: An augmented reality magic mirror system for anatomy education. In *2012 IEEE Virtual Reality Workshops* (pp. 115-116). IEEE.

Chaehan, S. (2021), Exploring Meta Learning: Parameterizing the Learning-to-learn Process for Image Classification. *International Conference on Artificial Intelligence in Information and Communication (ICAIIC).* IEEE.

Chaturvedi, S. S., Gupta, K., & Prasad, P. S. (2020). Skin lesion analyser: An efficient seven-way multi-class skin cancer classification using MobileNet. *Advances in Intelligent Systems and Computing*, 165–176. doi:10.1007/978-981-15-3383-9_15

Chawla, S., & Oberoi, A. (2011). *Algorithm for Iris Segmentation and Normalization using Hough Transform.* International Conference on Advanced Computing and Communication TechnologiesAt: Panipat, Haryana, India.

Chen, Y., Fan, R., & Yang, X. (2018). Extraction of Urban Water Bodies from High-Resolution Remote-Sensing Imagery Using Deep Learning. *Water, 10*, 585.

Chen, Z., & Tong, Y. (2017). *Face super-resolution through Wasserstein Gans.* arXiv preprint arXiv.02438.

Cheng, G., Han, J., & Lu, X. (2017). Remote sensing image scene classification: Benchmark and state of the art. *Proceedings of the IEEE, 105*(10), 1865–1883. doi:10.1109/JPROC.2017.2675998

Chen, H.-T., Wu, Y.-C., & Hsu, C.-C. (2016). Daytime Preceding Vehicle Brake Light Detection Using Monocular Vision. *IEEE Sensors Journal, 16*(1), 120–131. doi:10.1109/JSEN.2015.2477412

Chen, L. C., Zhu, Y., Papandreou, G., Schroff, F., & Adam, H. (2018). Encoder decoder with Atrous separable convolution for semantic image segmentation. *Proc. Eur. Conf. Comput. Vis. (ECCV),* (pp. 801–818). IEEE. 10.1007/978-3-030-01234-2_49

Chen, Y., Tai, Y., Liu, X., Shen, C., & Yang, J. (2018). FSRNet: End-to-End Learning Face Super-Resolution with Facial 23. Priors. In *Proceedings of the IEEE Computer Society Conference on Computer Vision and Pattern Recognition* (pp. 2492–2501). IEEE. 10.1109/CVPR.2018.00264

Coles, L. S. (1977). The application of artificial intelligence to medicine. *Futures, 9*(4), 315–323. doi:10.1016/0016-3287(77)90097-0

Cootes, T.F., Lindner, C., Carmona, I.T., & Carreira, M.J. (2019). Fully Automatic Teeth Segmentation in Adult OPG Images. In: Vrtovec, T., Yao, J., Zheng, G., Pozo, J. (eds) Computational Methods and Clinical Applications in Musculoskeletal Imaging. Springer, Cham.] doi:10.1007/978-3-030-11166-3_2

Cootes, T., Beeston, C., Edwards, G., & Taylor, C. (1999). Unified Framework for Atlas Matching Using Active Appearance Models. *Lecture Notes in Computer Science, 1613*, 322–333. doi:10.1007/3-540-48714-X_24

Correa Silva, A. (2020). Meta-Learning Applications in Digital Image Processing. *Proceedings of International Conference on Systems, Signals and Image Processing (IWSSIP),* (pp 19-20). IEEE.

Davis, L. S., Rosenfeld, A., & Weszka, J. S. (1975). Region extraction by averaging and thresholding [J]. *IEEE Transactions on Systems, Man, and Cybernetics, 1975*(3), 383–388. doi:10.1109/TSMC.1975.5408419

Debelee, T., Schwenker, F., Rahimeto, S., & Ashenafi, D. Y. (2019). Evaluation of modified adaptive k-means segmentation algorithm. *Computational Visual Media, 5*(4), 347–361. doi:10.100741095-019-0151-2

Deepthi, S. (2021, June). Sandeep Kumar, Dr. Suresh L. (2021), Detection and Classification of Objects in Satellite Images using Custom CNN [IJERT]. *International Journal of Engineering Research & Technology (Ahmedabad), 10*(06).

DeVries, T., & Ramachandram, D. (2017). *Skin Lesion Classification Using Deep Multi-Scale Convolutional Neural Networks*. https://arxiv.org/abs/1703.01402

Dias, T., Gonçalves, R., da Costa, R. L., Pereira, L. F., & Dias, Á. (2023). The impact of artificial intelligence on consumer behaviour and changes in business activity due to pandemic effects. *Human Technology, 19*(1), 121–148. doi:10.14254/1795-6889.2023.19-1.8

Diouf, D. (2019). *Convolutional Neural Network and decision support in medical imaging: case study of the recognition of blood cell subtypes*. arXiv preprint arXiv:1911.08010

Dogru, A. K., & Keskin, B. B. (2020). AI in operations management: Applications, challenges and opportunities. *Journal of Data. Information & Management, 2*, 67–74.

Du, Y., Zhang, Y., Ling, F., Wang, Q., Li, W., & Li, X. (2016). 'Water Bodies' Mapping from Sentinel-2 Imagery with Modified Normalized Difference Water Index at 10-m Spatial Resolution Produced by Sharpening the SWIR Band. *Remote Sensing, 8*(354), 1-19.

Duan, L., & Hu, X. (2019). Multiscale Refinement Network for Water-body Segmentation in High-Resolution Satellite Imagery. *IEEE Geoscience and Remote Sensing Letters, 17*(4), 686–690. doi:10.1109/LGRS.2019.2926412

Du, R. Y., Hu, Y., & Damangir, S. (2015). Leveraging trends in online searches for product features in market response modeling. *Journal of Marketing, 79*(1), 29–43. doi:10.1509/jm.12.0459

Du, X., Cai, Y., Wang, S., & Zhang, L. (2016). Overview of deep learning. *31st Youth Academic Annual Conference of Chinese Association of automation (YAC),* (pp.159–164). YAC. 10.1109/YAC.2016.7804882

El Shair, Z., & Rawashdeh, S. A. (2022). High-Temporal-Resolution Object Detection and Tracking Using Images and Events. *Journal of Imaging, 8*(8), 210. doi:10.3390/jimaging8080210 PMID:36005453

Fang, W., Wang, C., Chen, X., Wan, W., Li, H., Zhu, S., Fang, Y., Liu, B., & Hong, Y. (2019). Recognizing Global Reservoirs from Landsat 8 Images: A Deep Learning Approach. *IEEE Journal of Selected Topics in Applied Earth Observations and Remote Sensing, 12*(9), 3168–3177. doi:10.1109/JSTARS.2019.2929601

Fan, Z., Hu, X., Chen, C., Wang, X., & Peng, S. (2020, December). Facial image super-resolution guided by adaptive geometric features. *EURASIP Journal on Wireless Communications and Networking, 2020*(1), 149. doi:10.118613638-020-01760-y

Feng, M., Sexton, J. O., Channan, S., & Townshend, J. R. (2016). A global, high resolution (30-m) inland water body dataset for 2000: First results of a topographic–spectral classification algorithm. *International Journal of Digital Earth, 9*(2), 113–133. doi:10.1080/17538947.2015.1026420

Feng, W., Sui, H., Huang, W., Chuan, X., & Kaiquiang, A. (2019). Water body Extraction from Very High Resolution Remote Sensing Imagery using Deep U-Net and Superpixel Based Conditional Random Field Model. *IEEE Geoscience and Remote Sensing Letters, 16*(4), 618–622. doi:10.1109/LGRS.2018.2879492

Feng, Y., Duives, D. C., & Hoogendoorn, S. P. (2022, March 1). Development and evaluation of a VR research tool to study wayfinding behaviour in a multi-story building. *Safety Science, 147,* 105573. doi:10.1016/j.ssci.2021.105573

Feyisa, G. L., Meilby, H., Fensholt, R., & Proud, S. R. (2014). Automated water extraction index: A new technique for surface water mapping using Landsat imagery. *Remote Sensing of Environment, 140,* 23–35. doi:10.1016/j.rse.2013.08.029

Finn, C., Abbeel, P., & Levine, S. (2022). *Optimizing Learning: A Case Study in Meta-Learning for Computer Vision.* arXiv. (https://arxiv.org/abs/1703.03400)

Finn, C., Abbeel, P., & Levine, S. (2017). Model-agnostic meta-learning for fast adaptation of deep networks. In *International Conference on Machine Learning (ICML).* IEEE.

Finn, C., Xu, K., & Levine, S. (2019) *Probabilistic Model-Agnostic Meta-Learning. 32nd Conference on Neural Information Processing Systems (NeurIPS 2018),* Montréal, Canada

Frese, U., Wagner, R., & Röfer, T. (2010, September). A SLAM Overview from a User's Perspective. *Kunstliche Intelligenz, 24*(3), 191–198. doi:10.100713218-010-0040-4

Furkan, I., Bovik, A. C., & Paola, P. (2017). Surface Water mapping by Deep Learning. *IEEE Journal of Selected Topics in Applied Earth Observations and Remote Sensing, 10*(11), 4909–4918. doi:10.1109/JSTARS.2017.2735443

Furman, J., & Seamans, R. (2019). AI and the Economy. *Innovation Policy and the Economy, 19*(1), 161–191. doi:10.1086/699936

Fuster, A., Goldsmith-Pinkham, P., Ramadorai, T., & Walther, A. (2022). Predictably unequal? The effects of machine learning on credit markets. *The Journal of Finance, 77*(1), 5–47. doi:10.1111/jofi.13090

Galloway, C., & Swiatek, L. (2018). Public relations and artificial intelligence: It's not (just) about robots. *Public Relations Review, 44*(5), 734–740. doi:10.1016/j.pubrev.2018.10.008

Ganesan, P., & Sajiv, G. (2017). A comprehensive study of edge detection for image processing applications. *International Conference on Innovations in Information, Embedded and Communication Systems (ICIIECS),* (pp. 1-6). IEEE. 10.1109/ICIIECS.2017.8275968

Goodfellow, I., Pouget-Abadie, J., & Mirza, M. (2017). Generative adversarial nets. *Proc. Conf. and Workshop on Neural Information Processing Systems.*

Gordon, R. (2013). Skin cancer: An overview of epidemiology and risk factors. *Seminars in Oncology Nursing, 29*(3), 160–169. doi:10.1016/j.soncn.2013.06.002 PMID:23958214

Grant, E., Finn, C., Peterson, J., Abbott, J., Levine, S., Darrell, T., & Griffiths, T. (2017) Concept acquisition through meta-learning. In *NIPS Workshop on Cognitively Informed Artificial Intelligence.* IEEE.

Graves, A., Wayne, G., & Danihelka, I. (2014). Neural Turing Machines, Neural and Evolutionary Computing. arXiv. https://doi.org/ doi:10.48550/arXiv.1410.5401

Guo, H., He, G., Jiang, W., Yin, R., Yan, L., & Leng, W. (2020). A Multi-Scale Water Extraction Convolutional Neural Network (MWEN) Method for GaoFen-1 Remote Sensing Images. *ISPRS International Journal of Geo-Information, 9*(4), 189. doi:10.3390/ijgi9040189

Gupta, S., Panwar, A., Kapruwan, A., Chaube, N., & Chauhan, M. (2022, February). Real Time Analysis of Diabetic Retinopathy Lesions by Employing Deep Learning and Machine Learning Algorithms using Color Fundus Data. In *2022 International Conference on Innovative Trends in Information Technology (ICITIIT)* (pp. 1-5). IEEE. 10.1109/ICITIIT54346.2022.9744228

Habibzadeh, M., Krzyżak, A., & Fevens, T. (2013). White blood cell differential counts using convolutional neural networks for low resolution images. *International Conference on Artificial Intelligence and Soft Computing.* Springer, Berlin, Heidelberg. 10.1007/978-3-642-38610-7_25

Hamza-Lup, F. G., Santhanam, A. P., Imielinska, C., Meeks, S. L., & Rolland, J. P. (2007, January 2). Distributed augmented reality with 3-D lung dynamics—A planning tool concept. *IEEE Transactions on Information Technology in Biomedicine, 11*(1), 40–46. doi:10.1109/TITB.2006.880552 PMID:17249402

Harley, A. W. (2015). An interactive node-link visualization of Convolutional Neural Networks. *Advances in Visual Computing*, 867–877. doi:10.1007/978-3-319-27857-5_77

Hayat, K. (2018). Multimedia super-resolution via deep learning: A survey. *Digital Signal Processing: A Review Journal, 81.* . doi:10.1016/j.dsp.2018.07.005

Hospedales, T., Antoniou, A., Micaelli, P., & Storkey, A. (2022). Meta-Learning in Neural Networks: A Survey. *IEEE Transactions on Pattern Analysis and Machine Intelligence, 44*(9). PMID:33974543

Huaxiu. (2021). *A Survey on Meta-Learning. 35th Conference on Neural Information Processing Systems (NeurIPS 2021).* arXiv. (https://arxiv.org/abs/1810.03548)

Hugo. (2023). A Tutorial. *International conference on Artificial Intelligence.* Springer.

Huisman, M., van Rijn, J. N., & Plaat, A. (2021). A survey of deep meta-learning. *Artificial Intelligence Review, 54*(6), 4483–4541. doi:10.100710462-021-10004-4

ISIC Archive. (n.d.). *Archive.* ISIC. https://isic-archive.com/

Janjua, H. U., Akhtar, M., & Hussain, F. (2016). Effects of sugar, salt and distilled water on white blood cells and platelet cells: A review. *Zhong Liu Za Zhi, 4*(1), 354–358. doi:10.17554/j.issn.1819-6187.2016.04.73

Jaworek-Korjakowska, J., Kleczek, P., & Gorgon, M. (2019). Melanoma thickness prediction based on convolutional neural network with VGG-19 model transfer learning. In *2019 IEEE/CVF Conference on Computer Vision and Pattern Recognition Workshops (CVPRW).* IEEE. 10.1109/CVPRW.2019.00333

Jesorsky, O., Kirchberg, K. J., & Frischholz, R. W. (2001). Robust Face Detection Using the Hausdorff Distance. In J. Bigun & F. Smeraldi (Eds.), Lecture Notes in Computer Science: Vol. 2091. *Audio- and Video-Based Biometric Person Authentication. AVBPA 2001.* Springer. doi:10.1007/3-540-45344-X_14

Jeyakumar, J. P., Jude, A., Priya Henry, A. G., & Hemanth, J. (2022). Comparative analysis of melanoma classification using Deep Learning techniques on dermoscopy images. *Electronics (Basel), 11*(18), 2918. doi:10.3390/electronics11182918

Jha, S., & Topol, E. J. (2016). Adapting to artificial intelligence: Radiologists and pathologists as information specialists. *Journal of the American Medical Association, 316*(22), 2353–2354. doi:10.1001/jama.2016.17438 PMID:27898975

Jiang, J. (2016). Noise Robust Face Image Super-Resolution Through Smooth Sparse Representation. *IEEE Transactions on Cybernetics.* PMID:28113611

Ji, S. (2020). Kullback-Leibler Divergence Metric Learning. *IEEE Transactions on Cybernetics*. PMID:32721911

Jocher, G. (2020). "ultralytics/yolov5," *Github Repository*. YOLOv5.

Jones, S., & Dawkins, S. (2018). The sensorama revisited: evaluating the application of multi-sensory input on the sense of presence in 360-degree immersive film in virtual reality. In *Augmented reality and virtual reality* (pp. 183–197). Springer. doi:10.1007/978-3-319-64027-3_13

Kamphuis, C., Barsom, E., Schijven, M., & Christoph, N. (2014, September). Augmented reality in medical education? *Perspectives on Medical Education*, *3*(4), 300–311. doi:10.1007/S40037-013-0107-7 PMID:24464832

Kamra, P., Vishraj, R., & Kanica, S. G. (2015). Performance Comparison of Image Segmentation Techniques for Lung Nodule Detection in CT Images. *International Conference on Signal Processing, Computing and Control (ISPCC)*, (pp. 302-306). IEEE. 10.1109/ISPCC.2015.7375045

Kapoor, L., & Thakur, S. (2017). A Survey on Brain Tumor Detection Using Image Processing Techniques. *2017 7th International Conference on Cloud Computing, Data Science & Engineering – Confluence*. IEEE.

Kaushik, H., Singh, D., Kaur, M., Alshazly, H., Zaguia, A., & Hamam, H. (2021). Diabetic retinopathy diagnosis from fundus images using stacked generalization of deep models. *IEEE Access : Practical Innovations, Open Solutions*, *9*, 108276–108292. doi:10.1109/ACCESS.2021.3101142

Khan, I., Zhang, X., Rehman, M., & Ali, R. (2020). *A Literature survey and empirical study of Meta-Learning for Classifier Selection* (Vol. B). IEEE.

Khan, S., Tomar, S., Fatima, M., & Khan, M. Z. (2022). Impact of artificial intelligent and industry 4.0 based products on consumer behaviour characteristics: A meta-analysis-based review. *Sustainable Operations and Computers*, *3*, 218–225. doi:10.1016/j.susoc.2022.01.009

Khodak, M., Balcan, M. F., & Talwalkar, A. (2019) Provable Guarantees for Gradient-Based Meta-Learning. *Proceedings of the 36 th International Conference on Machine Learning, Long Beach, California*, 97. IEEE.

Khor, W. S., Baker, B., Amin, K., Chan, A., Patel, K., & Wong, J. (2016, December). Augmented and virtual reality in surgery—the digital surgical environment: Applications, limitations and legal pitfalls. *Annals of Translational Medicine*, *4*(23), 454. doi:10.21037/atm.2016.12.23 PMID:28090510

Kim, T., Lee, H., Kim, M. Y., Kim, S., & Duhachek, A. (2023). AI increases unethical consumer behavior due to reduced anticipatory guilt. *Journal of the Academy of Marketing Science*, *51*(4), 785–801. doi:10.100711747-021-00832-9

Komeil, R., Anuar, A., Karim, S., & Sharifeh, H. (2015). A new approach for surface water change detection: Integration of pixel level image fusion and image classification techniques. *International Journal of Applied Earth Observation and Geoinformation*, *34*, 226–234. doi:10.1016/j.jag.2014.08.014

Kong, H., Akakin, H. C., & Sarma, S. E. (2012). A Generalized Laplacian of Gaussian Filter for Blob Detection and Its Applications. *IEEE Transactions on Cybernetics*, *43*(6), 1719–1733. doi:10.1109/TSMCB.2012.2228639 PMID:23757570

Kopalle, P. K., Gangwar, M., Kaplan, A., Ramachandran, D., Reinartz, W., & Rindfleisch, A. (2022). Examining artificial intelligence (AI) technologies in marketing via a global lens: Current trends and future research opportunities. *International Journal of Research in Marketing*, *39*(2), 522–540. doi:10.1016/j.ijresmar.2021.11.002

Krueger, M. W., Gionfriddo, T., & Hinrichsen, K. (1985, April). VIDEOPLACE—an artificial reality. In *Proceedings of the SIGCHI conference on Human factors in computing systems* (pp. 35-40).

Kshitij, M., & Prasad, R. (2015). Automatic Extraction of Water Bodies from Landsat Imagery Using Perceptron Model. *Journal of Computational Environmental Sciences.*. doi:10.1155/2015/903465

Kuleto, V., Ilić, M., Dumangiu, M., Ranković, M., Martins, O. M., Păun, D., & Mihoreanu, L. (2021). Exploring opportunities and challenges of artificial intelligence and machine learning in higher education institutions. *Sustainability (Basel)*, *13*(18), 10424. doi:10.3390u131810424

Kumar, G., Rampavan, M., & Paul Ijjina, E. (2021). Deep Learning based Brake Light Detection for Two Wheelers. *2021 12th International Conference on Computing Communication and Networking Technologies (ICCCNT)*, (pp. 1-4). IEEE. 10.1109/ICCCNT51525.2021.9579918

Kuo, C., Cheng, S., Lin, C., Hsiao, K., & Lee, S. (2017). *Texture-based Treatment Prediction by Automatic Liver Tumor Segmentation on Computed Tomography*. IEEE.

Kurtosis. (2023). Hyperparameter Optimization in Meta-Learning. *Science direct, 260*.

Lahsen, M. (2020). Should AI be designed to save us from ourselves?: Artificial intelligence for sustainability. *IEEE Technology and Society Magazine*, *39*(2), 60–67. doi:10.1109/MTS.2020.2991502

Lee, M., & Park, J. S. (2022). Do parasocial relationships and the quality of communication with AI shopping chatbots determine middle-aged women consumers' continuance usage intentions? *Journal of Consumer Behaviour*, *21*(4), 842–854. doi:10.1002/cb.2043

Lee, T.-H., Chou, H.-S., Chen, T.-Y., Lo, W.-S., Zhang, J.-T., Chen, C.-A., Lin, T.-L., & Chen, S.-L. (2022). Laplacian of Gaussian Based on Color Constancy Algorithm for Surrounding Image Stitching Application. *IEEE International Conference on Consumer Electronics*, (pp. 287-288). IEEE. 10.1109/ICCE-Taiwan55306.2022.9869055

Li T., Su, Xin., Liu, Su., Liang, W., Hsieh, M.Y., Chen, Z., Liu, X.C., & Zhang, H. (2022) *Memory-augmented meta-learning on meta-path for fast adaptation cold-start recommendation*. Taylor and Francis.

Li, Z., Zhou, F., Chen, F., & Li, H. (2017) Meta-SGD: Learning to learn quickly for few-shot learning. arXiv

Li, F., Yan, L., Wang, Y., Shi, J., Chen, H., Zhang, X., Jiang, M., Wu, Z., & Zhou, K. (2020). Deep learning-based automated detection of glaucomatous optic neuropathy on color fundus photographs. *Graefe's Archive for Clinical and Experimental Ophthalmology, 258*(4), 851–867. doi:10.100700417-020-04609-8 PMID:31989285

Li, J., Zhou, Y., Ding, J., Chen, C., & Yang, X. (2020). ID Preserving Face Super-Resolution Generative Adversarial Networks. *IEEE Access : Practical Innovations, Open Solutions, 8*, 1. doi:10.1109/ACCESS.2020.3011699

Limna, P. (2022). Artificial Intelligence (AI) in the hospitality industry: A review article. *Int. J. Comput. Sci. Res, 6*, 1–12.

Lin, T.-Y. (2014). Microsoft coco: Common objects in context. *European conference on computer vision.* Springer, Cham. 10.1007/978-3-319-10602-1_48

Lin, H., Shi, Z., & Zou, Z. (2017). 'Maritime Semantic Labeling of Optical Remote Sensing images with multiscale fully convolutional networks. *Remote Sensing (Basel), 9*(5), 480. doi:10.3390/rs9050480

Lin, P. H., Wooders, A., Wang, J. T. Y., & Yuan, W. M. (2018). Artificial intelligence, the missing piece of online education? *IEEE Engineering Management Review, 46*(3), 25–28. doi:10.1109/EMR.2018.2868068

Liu, Y., Shi1, C., Lin, B., Ha, C., Papanikolaou, N. (2009). Delivery of four-dimensional radiotherapy with Track Beam for moving target using an AccuKnife dual-layer MLC: Dynamic phantoms study. *Journal of Applied Clinical Medical Physics, 10*(2), 2926. doi:10.1120/jacmp. v10i2.2926 PMID:19458594

Liu, T. Z., Yu, J., & Tang, F. (2018, August). J.Deep Meta-learning in Recommendation Systems. *Survey (London, England), 37*(4), 11.

Liu, X. H., Wang, T., Lin, J. P., & Wu, M. B. (2018, December 2). Using virtual reality for drug discovery: A promising new outlet for novel leads. *Expert Opinion on Drug Discovery, 13*(12), 1103–1114. doi:10.1080/17460441.2018.1546286 PMID:30457399

Long, Y., Wang, Z., Tian, S., Ye, F., Ding, J., & Jun, K. (2017). Convolutional Neural Networks for Water-body Extraction from Landsat Imagery. *International Journal of Computational Intelligence and Applications, 16*(1).

Luo, X., Tong, S., Fang, Z., & Qu, Z. (2019). Frontiers: Machines vs. humans: The impact of artificial intelligence chatbot disclosure on customer purchases. *Marketing Science, 38*(6), 937–947. doi:10.1287/mksc.2019.1192

Mahmoud, F. & Al-Ahmad, H. (2016). Two dimensional filters for enhancing the resolution of interpolated CT scan images. *2016 12th International Conference on Innovations in Information Technology (IIT),* (pp. 1-6). IEEE. 10.1109/INNOVATIONS.2016.7880034

Mane, S., & Shinde, S. (2018). A method for melanoma skin cancer detection USING Dermoscopy Images. In *2018 Fourth International Conference on Computing Communication Control and Automation (ICCUBEA)*. IEEE. 10.1109/ICCUBEA.2018.8697804

Mariani, M. M., Perez-Vega, R., & Wirtz, J. (2022). AI in marketing, consumer research and psychology: A systematic literature review and research agenda. *Psychology and Marketing*, *39*(4), 755–776. doi:10.1002/mar.21619

Marks, R. (1995). An overview of skin cancers. *Cancer*, *75*(S2), 607–612. doi:10.1002/1097-0142(19950115)75:2+<607::AID-CNCR2820751402>3.0.CO;2-8 PMID:7804986

McFeeters, S. K. (1996). The use of the normalized difference water index (NDWI) in the delineation of open water features. *International Journal of Remote Sensing*, *17*(7), 1425–1432. doi:10.1080/01431169608948714

Miao, Z., Kun, F., Hao, S., Xian, S., & Yan, M. (2018). Automatic Water-body Segmentation From High Resolution Setellite Images via Deep Networks. *IEEE Geoscience and Remote Sensing Letters*, *15*(4), 602–606. doi:10.1109/LGRS.2018.2794545

Milgram, P., Takemura, H., Utsumi, A., & Kishino, F. (1995). Augmented reality: A class of displays on the reality-virtuality continuum. In *Telemanipulator and telepresence technologies*, *2351*, 282-292. Spie.

Mina, T., & Wathiq, M. (2018). Detection of Water-Bodies Using Semantic Segmentation. *International Conference on Signal Processing and Information Security (ICSPIS)*. IEEE.

Mind, D. S.-D. (n.d.). *Umělá inteligence, machine learning a neuronové sítě*. Data Science - Data Mind. http://www.datamind.cz/cz/Sluzby-Data-Science/umela-intelingence-AI-ML-machine-learning-neural-net

Moghbel, M., Mashohor, S., Mahmud, R., & Iqbal Bin Saripan, M. (2016). Automatic liver tumor segmentation on computed tomography for patient treatment planning and monitoring. *EXCLI Journal*, *15*, 406–423. PMID:27540353

Monostori, L. (2003). AI and machine learning techniques for managing complexity, changes and uncertainties in manufacturing. *Engineering Applications of Artificial Intelligence*, *16*(4), 277–291. doi:10.1016/S0952-1976(03)00078-2

Mugahed, A. (2018). A fully integrated computer-aided diagnosis system for digital X-ray mammograms via deep learning detection, segmentation, and classification. *International Journal of Medical Informatics*, *117*, 44–54. doi:10.1016/j.ijmedinf.2018.06.003 PMID:30032964

Muhadi, N. A., Abdullah, A. F., Bejo, S. K., Mahadi, M. R., & Mijic, A. (1825). 2020, 'Image Segmentation Methods for Flood Monitoring System', Water. *An MDPI Journal*, *12*, 1–10.

Mule, D. B., Chowhan, S. S., & Somwanshi, D. R. (2019). Detection and Classification of Non-proliferative Diabetic Retinopathy Using Retinal Images. In K. Santosh & R. Hegadi (Eds.), *Recent Trends in Image Processing and Pattern Recognition. RTIP2R 2018. Communications in Computer and Information Science* (Vol. 1036). Springer.

Muller, E., & Peres, R. (2019). The effect of social networks structure on innovation performance: A review and directions for research. *International Journal of Research in Marketing, 36*(1), 3–19. doi:10.1016/j.ijresmar.2018.05.003

Mun, J., & Kim, J. (2020, November). Universal super-resolution for face and non-face regions via a facial feature network. *Signal, Image and Video Processing, 14*(8), 1601–1608. doi:10.100711760-020-01706-3

Murthy, T. S. D., & Sadashivappa, G. (2014). Brain tumor segmentation using thresholding, morphological operations and extraction of features of tumor. *2014 International Conference on Advances in Electronics Computers and Communications*. IEEE. 10.1109/ICAECC.2014.7002427

Muthukrishnan, R., & Radha, M. (2011). Edge Detection Techniques For Image Segmentation [IJCSIT]. *International Journal of Computer Science and Information Technologies, 3*(6).

Nasr-Esfahani, E., Samavi, S., Karimi, N., Soroushmehr, S. M. R., Jafari, M. H., Ward, K., & Najarian, K. (2016). Melanoma Detection by Analysis of Clinical Images Using Convolutional Neural Network. In *Proceedings of the 2016 38th Annual International Conference of the IEEE Engineering in Medicine and Biology Society (EMBC)*, (pp. 1373). IEEE. 10.1109/EMBC.2016.7590963

Ngai, E. W., Xiu, L., & Chau, D. C. (2009). Application of data mining techniques in customer relationship management: A literature review and classification. *Expert Systems with Applications, 36*(2), 2592–2602. doi:10.1016/j.eswa.2008.02.021

Nguyen, B. P., Heemskerk, H., So, P. T., & Tucker-Kellogg, L. (2016). Superpixel-based segmentation of muscle fibers in multi-channel microscopy. *BMC Systems Biology, 10*(S5, Suppl 5), 124. doi:10.118612918-016-0372-2 PMID:28105947

Nguyen, T. M., Quach, S., & Thaichon, P. (2022). The effect of AI quality on customer experience and brand relationship. *Journal of Consumer Behaviour, 21*(3), 481–493. doi:10.1002/cb.1974

Nilanchal, PRohit, M. (2014). Extraction of Impervious Features from Spectral indices using Artificial Neural Network. *Arabian Journal of Geosciences, 8*(6).

Nobuyuki, O. (1979). A Threshold Selection Method from Gray-Level Histograms. *IEEE Transactions on Systems, Man, and Cybernetics, SMC-9*(1), 62–66.

Noppitak, S., Gonwirat, S., & Surinta, O. (2020). *Instance Segmentation of Water-body from Aerial Image using Mask Region based Convolutional Neural Network*. The 3rd International Conference on Information Science and Systems (ICISS), Cambridge, UK.

Olmedo, H. (2013, January 1). Virtuality continuum's state of the art. *Procedia Computer Science, 25*, 261–270. doi:10.1016/j.procs.2013.11.032

Omarov, B., Tursynbayev, A., Zhakypbekova, G., & Beissenova, G. (2022). Effect Of Chatbots In Digital Marketing To Perceive The Consumer Behaviour. *Journal of Positive School Psychology, 6*(8), 7143–7161.

Özyurt, F. (2020). A fused CNN model for WBC detection with MRMR feature selection and extreme learning machine. *Soft Computing*, *24*(11), 8163–8172. doi:10.100700500-019-04383-8

Panch, T., Szolovits, P., & Atun, R. (2018). Artificial intelligence, machine learning and health systems. *Journal of Global Health*, *8*(2), 020303. doi:10.7189/jogh.08.020303 PMID:30405904

Parihar, V., & Yadav, S. (2021). Comparison estimation of effective consumer future preferences with the application of AI. *Vivekananda Journal of Research*, *10*, 133–145.

Parisay, M., Poullis, C., & Kersten, M. (2020). Eyetap: A novel technique using voice inputs to address the midas touch problem for gaze-based interactions. arXiv preprint arXiv:2002.08455. Feb 19.

Park, H., Lee, G., Kim, S., Ryu, G., Jeong, A., Park, S., & Sagong, M. (2021) A Meta-Learning Approach for Medical Image Registration. *IEEE 19th International Symposium on Biomedical Imaging (ISBI)*. IEEE.

Park, M., Jung, J., Oh, Y., & You, H. (2010). Assessment of Epicardial Fat Volume With Threshold-Based 3-Dimensional Segmentation in CT: Comparison With the 2-Dimensional Short Axis-Based Method. *Korean Circulation Journal*. .] doi:10.4070/kcj.2010.40.7.328

Pham-Duc, B., Prigent, C., & Aires, F. (2017). Surface water monitoring within Cambodia and the Vietnamese Mekong Delta over a year, with Sentinel-1 SAR observations. *Water, 9*(6), 366

Pirhonen, J., Ojala, R., Kivekäs, K., Vepsäläinen, J., & Tammi, K. (2022). Brake Light Detection Algorithm for Predictive Braking. *Applied Sciences (Basel, Switzerland)*, *12*(6), 2804. doi:10.3390/app12062804

Prabhu. (2019, November 21). Understanding of convolutional neural network (CNN) - deep learning. *Medium*. https://medium.com/@RaghavPrabhu/understanding-of-convolutional-neural-network-cnn-deep-learning-99760835f148

Priesnitz, J., Rathgeb, C., Buchmann, N., Busch, C., & Margraf, M. (2021). Buchmann (2021). An overview of touchless 2D fingerprint recognition. *EURASIP Journal on Image and Video Processing*, *8*(1), 8. doi:10.118613640-021-00548-4

PS Staff. (2016). *What Is the Difference between Melanoma And non-Melanoma Skin Cancer?* PSS. https://www.premiersurgical.com/01/whats-the-difference-between-melanoma-and-non-melanoma-skin-cancer/

Pujara, A. (2020, July 15). Image classification with MobileNet. *Medium*. https://medium.com/analytics-vidhya/image-classification-with-mobilenet-cc6fbb2cd470

Qi, G.-J. 'Loss-sensitive generative adversarial networks on Lipschitz densities', arXiv preprint arXiv.06264, 2017.

Qiao, L., Zhu, Y., & Zhou, H.WebQiao. (2020). Diabetic Retinopathy Detection Using Prognosis of Microaneurysm and Early Diagnosis System for Non-Proliferative Diabetic Retinopathy Based on Deep Learning Algorithms. *IEEE Access : Practical Innovations, Open Solutions, 8,* 104292–104302. doi:10.1109/ACCESS.2020.2993937

Qureshi, I., Ma, J., & Abbas, Q. (2021). Diabetic retinopathy detection and stage classification in eye fundus images using active deep learning. *Multimedia Tools and Applications, 80*(8), 11691–11721. doi:10.100711042-020-10238-4

Rabby, F., Chimhundu, R., & Hassan, R. (2021). Artificial intelligence in digital marketing influences consumer behaviour: A review and theoretical foundation for future research. *Academy of Marketing Studies Journal, 25*(5), 1–7.

Ratamero, E. M., Bellini, D., Dowson, C. G., & Römer, R. A. (2018, June). Touching proteins with virtual bare hands. *Journal of Computer-Aided Molecular Design, 32*(6), 703–709. doi:10.100710822-018-0123-0 PMID:29882064

Ravi, S., & Larochelle, H. (2017) Optimization as a Model for Few-Shot Learning. *International Conference on Learning* . IEEE

Razzak, I., Shoukat, G., Naz, S., & Khan, T. M. (2020). Skin lesion analysis toward accurate detection of melanoma using multistage fully connected residual network. In *2020 International Joint Conference on Neural Networks (IJCNN).* IEEE. 10.1109/IJCNN48605.2020.9206881

Redmon, J. (2016). You only look once: Unified, real-time object detection. *Proceedings of the IEEE conference on computer vision and pattern recognition.* IEEE. 10.1109/CVPR.2016.91

Ren, Y., Shi, B., Hou, R., Grimm, L., Mazurowski, M. A., Marks, J., King, L., Maley, C., Hwang, S., & Lo, J. (2018). Learning better deep features for the prediction of occult invasive disease in ductal carcinoma in situ through transfer learning. In *Medical Imaging 2018: Computer-Aided Diagnosis.* IEEE. doi:10.1117/12.2293594

Richardson, A., Bracegirdle, L., McLachlan, S. I., & Chapman, S. R. (2013, February 12). Use of a three-dimensional virtual environment to teach drug-receptor interactions. *American Journal of Pharmaceutical Education, 77*(1), 11. doi:10.5688/ajpe77111 PMID:23459131

Safia, D. (2016). *Batouche, M.* Quantum Genetic Computing and Cellular Automata for Solving Edge Detection.

Sanchez, G., Dalmau, O., & Alarcon, T. (2018). Selection and Fusion of Spectral Indices to Improve Water Body Discrimination. *IEEE Access, 6,* 72952-72961.

Santoro, A., Bartunov, S., Botvinick, M., Wierstra, D., & Lillicrap, T. (2016). *Meta-Learning with Memory-Augmented Neural Networks.* Proceedings of the 33rd International Conference on Machine Learning, New York, NY.

Santos, J., Pyrcz, M., & Prodanović, M. (2022). 3D Dataset of binary images: A collection of synthetically created digital rock images of complex media. *Data in Brief, 40.* doi:10.1016/j.dib.2022.107797

Sarki, R., Ahmed, K., Wang, H., & Zhang, Y. (2020). Automatic detection of diabetic eye disease through deep learning using fundus images: A survey. *IEEE Access : Practical Innovations, Open Solutions, 8*, 151133–151149. doi:10.1109/ACCESS.2020.3015258

Sathesh, A., Eisa, E., & Babikir, A. (2021), Hybrid Parallel Image Processing Algorithm for Binary Images with Image Thinning Technique. *Journal of Artificial Intelligence and Capsule Networks, 03*(3), 243-258. . doi:10.36548/jaicn.2021.3.007

Selvakumar, P., & Hariganesh, S. (2016). The performance analysis of edge detection algorithms for image processing. *International Conference on Computing Technologies and Intelligent Data Engineering (ICCTIDE'16),* (pp. 1-5). IEEE. 10.1109/ICCTIDE.2016.7725371

Selvaraj, M. G., Vergara, A., Ruiz, H., Safari, N., Elayabalan, S., Ocimati, W., & Blomme, G. (2019). AI-powered banana diseases and pest detection. *Plant Methods, 15*(1), 92. doi:10.118613007-019-0475-z

Sesikala, B., Harikiran, J., & Sai Chandana, B. (2022, April). A Study on Diabetic Retinopathy Detection, Segmentation and Classification using Deep and Machine Learning Techniques. In *2022 6th International Conference on Trends in Electronics and Informatics (ICOEI)* (pp. 1419-1424). IEEE.

Shafiee, M. J. (2017). *Fast YOLO: A fast you only look once system for real-time embedded object detection in video.* arXiv preprint arXiv:1709.05943 ().

Shah, A., Kulkarni, J., Patil, R., Sisode, V., & Gogave, S. (2020). Traffic Control System and Technologies: A Survey. *International Journal of Engineering and Technical Research, 9*(01). doi: . doi:0.17577/IJERTV9IS010246

Shanmathi, N., & Jagannath, M. (2018). Computerised Decision Support System for Remote Health Monitoring: A Systematic Review. *IRBM, 39*(5), 359-367.] doi:10.1016/j.irbm.2018.09.007

Sharma, N. (2014, June). Image Segmentation and Medical Diagnosis. *International Journal of Engineering Trends and Technology, 12*(2), 94–97. doi:10.14445/22315381/IJETT-V12P216

Sharma, P. K. (2016). Dilation of Chisini-Jensen-Shannon Divergence. In *IEEE International Conference on Data Science and Advanced Analytics (DSAA).* IEEE.

Shelhamer, E., Long, J., & Darrell, T. (2017). Fully convolutional networks for semantic segmentation. *IEEE Transactions on Pattern Analysis and Machine Intelligence, 39*(4), 640–651. doi:10.1109/TPAMI.2016.2572683 PMID:27244717

Shinde, S., Kothari, A., & Gupta, V. (2018). YOLO based human action recognition and localization. *Procedia Computer Science, 133*, 831–838. doi:10.1016/j.procs.2018.07.112

Shoukat, A., Akbar, S., Hassan, S. A. E., Rehman, A., & Ayesha, N. (2021). An automated deep learning approach to diagnose glaucoma using retinal fundus images. In 2021 international conference on frontiers of information technology (FIT). IEEE.

Singh, V., Sharma, N., & Singh, S. (2020). A review of imaging techniques for plant disease detection. *Artificial Intelligence in Agriculture*. https://doi.org/ 4, 229-242. doi:10.1016/j.aiia.2020.10.002.Volume

Singh, L. K., Khanna, M., & Thawkar, S. (2022). A novel hybrid robust architecture for automatic screening of glaucoma using fundus photos, built on feature selection and machine learning-nature driven computing. *Expert Systems: International Journal of Knowledge Engineering and Neural Networks*, *39*(10), e13069. doi:10.1111/exsy.13069

Skarbez, R., Smith, M., & Whitton, M. C. (2021, March 24). Revisiting milgram and kishino's reality-virtuality continuum. *Frontiers in Virtual Reality*, *2*, 647997. doi:10.3389/frvir.2021.647997

Solbo, S., Malnes, E., Guneriussen, T., Solheim, I., & Eltoft, T. (2003). Mapping surface-water with Radarsat at arbitrary incidence angles. *Geoscience and Remote Sensing Symposium, IGARSS'03 Proceedings*. IEEE International. 10.1109/IGARSS.2003.1294494

Soler, L., Nicolau, S., Pessaux, P., Mutter, D., & Marescaux, J. (2014, April). Real-time 3D image reconstruction guidance in liver resection surgery. *Hepatobiliary Surgery and Nutrition*, *3*(2), 73. PMID:24812598

Soltani-Fesaghandis, G., & Pooya, A. (2018). Design of an artificial intelligence system for predicting success of new product development and selecting proper market-product strategy in the food industry. *The International Food and Agribusiness Management Review*, *21*(7), 847–864. doi:10.22434/IFAMR2017.0033

Speicher, M., Hall, B. D., & Nebeling, M. (2019). What is mixed reality? In Proceedings of the CHI conference on human factors in computing systems. ACM. doi:10.1145/3290605.3300767

Suganyadevi, S., Renukadevi, K., Balasamy, K., & Jeevitha, P. (2022, February). Diabetic Retinopathy Detection Using Deep Learning Methods. In *2022 First International Conference on Electrical, Electronics, Information and Communication Technologies (ICEEICT)* (pp. 1-6). IEEE. 10.1109/ICEEICT53079.2022.9768544

Sun, Z. (2022). Application of Image Super-Resolution Reconstruction in Gymnastics Training by Using Internet of Things Technology. Computational Intelligence and Neuroscience. https://doi.org/ doi:10.1155/2022/8133187

Sun, J., & Li, Y. (2021). MetaSeg: A survey of meta learning for image segmentation. *Cognitive Robotics.*, *1*, 83–91. doi:10.1016/j.cogr.2021.06.003

Sun, Y., Ren, Z., & Zheng, W. (2022). Research on Face Recognition Algorithm Based on Image Processing. *Computational Intelligence and Neuroscience*, *2022*, 9224203. Advance online publication. doi:10.1155/2022/9224203 PMID:35341202

Suzuki, K. (2017). Overview of deep learning in medical imaging. *Radiological Physics and Technology*, *10*(3), 257–273. doi:10.100712194-017-0406-5 PMID:28689314

Tahir, A., Munawar, H. S., Akram, J., Adil, M., Ali, S., Kouzani, A. Z., & Mahmud, M. (2022). M.A.P. Automatic Target Detection from Satellite Imagery Using Machine Learning. *Sensors (Basel)*, *22*(3), 1147. doi:10.339022031147 PMID:35161892

Tanzi, L., Piazzolla, P., Porpiglia, F., & Vezzetti, E. (2021). Real-time deep learning semantic segmentation during intra-operative surgery for 3D augmented reality assistance. *International Journal of Computer Assisted Radiology and Surgery*, *16*(9), 1435–1445. doi:10.100711548-021-02432-y PMID:34165672

Teleaba, F., Popescu, S., Olaru, M., & Pitic, D. (2021). Risks of observable and unobservable biases in artificial intelligence used for predicting consumer choice. *Amfiteatru Economic*, *23*(56), 102–119. doi:10.24818/EA/2021/56/102

Teoh, K. H., Ismail, R. C., Naziri, S. Z. M., Hussin, R., Isa, M. N. M., & Basir, M. S. S. M. (2021). "Face Recognition and Identification using Deep Learning Approach" Journal of Physics: Conference Series PAPER, OPEN ACCESS, KH Teoh. *Journal of Physics: Conference Series*, *1755*(1), 012006. doi:10.1088/1742-6596/1755/1/012006

Vatathanavaro, S., Tungjitnob, S., & Pasupa, K. (2018). White blood cell classification: a comparison between VGG-16 and ResNet-50 models. *Proceeding of the 6th joint symposium on computational intelligence (JSCI6)*. IEEE.

Ventola, C. L. (2019, May). Virtual reality in pharmacy: Opportunities for clinical, research, and educational applications. *P&T*, *44*(5), 267. PMID:31080335

Wagner, D. N. (2020). Economic patterns in a world with artificial intelligence. *Evolutionary and Institutional Economics Review*, *17*(1), 111–131. doi:10.100740844-019-00157-x

Wang, L., Chen, W., Yang, W., Bi, F., & Yu, F. R. (2020). A State-of-the-Art Review on Image Synthesis with Generative Adversarial Networks. In IEEE Access (Vol. 8). Institute of Electrical and Electronics Engineers Inc. doi:10.1109/ACCESS.2020.2982224

Wang, Q., Chen, M., Nie, F., & Li, X. (2020, January 1). Detecting coherent groups in crowd scenes by multiview clustering. *IEEE Transactions on Pattern Analysis and Machine Intelligence*, *42*(1), 46–58. Advance online publication. doi:10.1109/TPAMI.2018.2875002 PMID:30307858

Weber, F. D., & Schutte, R. (2019). State-of-the-art and adoption of artificial intelligence in retailing. *Digital Policy. Regulation & Governance*, *21*(3), 264–279. doi:10.1108/DPRG-09-2018-0050

Wirth, N. (2018). Hello marketing, what can artificial intelligence help you with? *International Journal of Market Research*, *60*(5), 435–438. doi:10.1177/1470785318776841

Wu, J., Liu, X., & Chen, S. (2023). Hyperparameter optimization through context-based meta-reinforcement learning with task-aware representation. *Knowledge-Based Systems, 260*. https://doi.org/ doi:10.1016/j.knosys.2022.110160

Wu, Z. & Leahy, R. (1993). An optimal graph theoretic approach to data clustering: Theory and its application to image segmentation [J]. *IEEE transactions on pattern analysis and machine intelligence, 1993, 15*(11), 1101- 1113.

Wu, W. (2016), Paralleled Laplacian of Gaussian (LoG) edge detection algorithm by using GPU. *Eighth International Conference on Digital Image Processing*. SPIE. 10.1117/12.2244599

Xu, J. (2017, August 15). An intuitive guide to deep network architectures. *Medium*. https://towardsdatascience.com/an-intuitive-guide-to-deep-network-architectures-65fdc477db41

Xu, X., & Porikli, F. (2017). Hallucinating very low-resolution unaligned and noisy face images by transformative discriminative autoencoders. *Proc. Int. Conf. on Computer Vision and Pattern Recognition.*

Xu, H. (2006). Modification of normalised difference water index (MNDWI) to enhance open water features in remotely sensed imagery. *International Journal of Remote Sensing*, *27*(14), 3025–3033. doi:10.1080/01431160600589179

Ya'nan, Z., Luo, J., Shen, Z., Hu, X., & Yang, H. (2014). Multiscale Water Body Extraction in Urban Environments from Satellite Images. *IEEE Journal of Selected Topics in Applied Earth Observations and Remote Sensing*, *7*(10), 4301–4312. doi:10.1109/JSTARS.2014.2360436

Yaman, O., & Tuncer, T. (2022). Exemplar pyramid deep feature extraction based cervical cancer image classification model using pap-smear images. *Biomedical Signal Processing and Control*, *73*, 103428. doi:10.1016/j.bspc.2021.103428

Yang, W., Zhang, X., Tian, Y., Wang, W., Xue, J., & Liao, Q. (2019). Deep Learning for Single Image Super-Resolution: A Brief Review. *IEEE Transactions on Multimedia*. doi:10.1109/TMM.2019.2919431

Yang, G., & Xu, F. (2011). Research and analysis of Image edge detection algorithm Based on the MATLAB. *Procedia Engineering*, *15*, 1313–1318. doi:10.1016/j.proeng.2011.08.243

YangL.LiuC. (2020). HiFaceGAN: Face Renovation via Collaborative Suppression and Replenishment. doi:10.1145/3394171.3413965

Yang, X., Qin, Q., Grussenmeyer, P., & Koehl, M. (2018). Urban surface water body detection with suppressed built-up noise based on water indices from Sentinel-2 MSI imagery. *Remote Sensing of Environment*, *219*, 259–270. doi:10.1016/j.rse.2018.09.016

Yao, H., Huang, L. K., Zhang, L., Wei, Y., Tian, L., Zou, J., & Huang, J. (2021) Improving generalization in meta-learning via task augmentation. In *Proceedings of the International Conference on Machine Learning, Virtual Event*. IEEE.

Ye, J., H., Chao, W., (2022) How to train your MAML to excel in Few-Shot Classification. *The International Conference on Learning Representations (ICLR)*. IEEE.

Yeom, S. (2011). *Augmented Reality for Learning Anatomy*. University of Tasmania.

Yudie, W., Zhiwei, L., Chao, Z., Xia, G., & Shen, H. (2019). Extracting Urban Water by Combining Deep Learning and Google Earth Engine. *IEEE journal of selected topics in applied earth observations & Remote Sensing*, *13*, 768-781.

Yu, K. (2019). ESRGAN: Enhanced Super-Resolution Generative Adversarial Networks. In *European Conference on Computer Vision*. Springer, .

Yu, X., Fernando, B., & Hartley, R. (2018). Super-resolving very low-resolution face images with supplementary attributes. *Proc. of the IEEE Conf. on Computer Vision and Pattern Recognition*. 10.1109/CVPR.2018.00101

Yu, X., & Porikli, F. (2016). Ultra-resolving face images by discriminative generative networks. *European Conf. on Computer Vision*, Amsterdam, Netherlands. 10.1007/978-3-319-46454-1_20

Zeiler, M. D., & Fergus, R. (2013). *Visualizing and Understanding Convolutional Networks*. https://arxiv.org/abs/1311.2901

Zhang, Y. (2019). *Efficient Hyperparameter Optimization for Deep Meta-Learning*. arXiv. (https://arxiv.org/abs/2103.09268)

Zhang. (2018). Super-identity convolutional neural network for face hallucination. *Proc. Eur. Conf. Comput. Vis. (ECCV)*. IEEE. 10.1007/978-3-030-01252-6_12

Zhang, H., Wang, P., Zhang, C., & Jiang, Z. (2019, July). A comparable study of CNN-based single image super-resolution for space-based imaging sensors. *Sensors (Basel)*, *19*(14), 3234. doi:10.339019143234 PMID:31340511

Zhang, Z., Zhang, X., Peng, C., Xue, X., & Sun, J. (2018). ExFuse: Enhancing feature fusion for semantic segmentation. *Proc. Eur. Conf. Comput. Vis. (ECCV)*, (pp. 269–284). IEEE. 10.1007/978-3-030-01249-6_17

Zhao, J., Zhang, M., Zhou, Z., Chu, J., & Cao, F. (2017). Automatic detection and classification of leukocytes using convolutional neural networks.(. *Medical & Biological Engineering & Computing*, *55*(8), 1287–1301. https://github.com/Shenggan/BCCD_Dataset. doi:10.100711517-016-1590-x PMID:27822698

Zheng, F., & Shao, L. (2018). A winner-take-All strategy for improved object tracking. *IEEE Transactions on Image Processing*, *27*(9), 4302–4313. doi:10.1109/TIP.2018.2832462 PMID:29870349

Zhenguo. (2023). Meta-Learning: A Survey. Science direct, 15.

Zheng, Y., Yang, C., & Merkulov, A. (2018). Breast cancer screening using convolutional neural network and follow-up Digital Mammography. *Computational Imaging*, *III*, 4. doi:10.1117/12.2304564

Zhi-Song, L. (2019). *Reference Based Face Super-Resolution*. IEEE.

Ziheng, S., Liping D., & Hui, F. (2019). Using long short-term memory recurrent neural network in land cover classification on Landsat and Cropland data layer time series. *International Journal of Remote Sensing, 40*(2), 593-614. doi:10.1080/01431161.2018.1516313

Zimeras, S. (2012). Segmentation Techniques of Anatomical Structures with Application in Radiotherapy Treatment Planning: Modern Practices in Radiation Therapy. *InTech*. doi:10.5772/34955

Zimeras, S. (2010). Segmentation Techniques of Anatomical Structures with Application in Radiotherapy Treatment Planning. *ACM Transactions on Graphics*, *29*(5), 134. doi:10.1145/1857907.1857910

Related References

To continue our tradition of advancing academic research, we have compiled a list of recommended IGI Global readings. These references will provide additional information and guidance to further enrich your knowledge and assist you with your own research and future publications.

Abbasnejad, B., Moeinzadeh, S., Ahankoob, A., & Wong, P. S. (2021). The Role of Collaboration in the Implementation of BIM-Enabled Projects. In J. Underwood & M. Shelbourn (Eds.), *Handbook of Research on Driving Transformational Change in the Digital Built Environment* (pp. 27–62). IGI Global. https://doi.org/10.4018/978-1-7998-6600-8.ch002

Abdulrahman, K. O., Mahamood, R. M., & Akinlabi, E. T. (2022). Additive Manufacturing (AM): Processing Technique for Lightweight Alloys and Composite Material. In K. Kumar, B. Babu, & J. Davim (Ed.), *Handbook of Research on Advancements in the Processing, Characterization, and Application of Lightweight Materials* (pp. 27-48). IGI Global. https://doi.org/10.4018/978-1-7998-7864-3.ch002

Agrawal, R., Sharma, P., & Saxena, A. (2021). A Diamond Cut Leather Substrate Antenna for BAN (Body Area Network) Application. In V. Singh, V. Dubey, A. Saxena, R. Tiwari, & H. Sharma (Eds.), *Emerging Materials and Advanced Designs for Wearable Antennas* (pp. 54–59). IGI Global. https://doi.org/10.4018/978-1-7998-7611-3.ch004

Ahmad, F., Al-Ammar, E. A., & Alsaidan, I. (2022). Battery Swapping Station: A Potential Solution to Address the Limitations of EV Charging Infrastructure. In M. Alam, R. Pillai, & N. Murugesan (Eds.), *Developing Charging Infrastructure and Technologies for Electric Vehicles* (pp. 195–207). IGI Global. doi:10.4018/978-1-7998-6858-3.ch010

Aikhuele, D. (2018). A Study of Product Development Engineering and Design Reliability Concerns. *International Journal of Applied Industrial Engineering*, 5(1), 79–89. doi:10.4018/IJAIE.2018010105

Al-Khatri, H., & Al-Atrash, F. (2021). Occupants' Habits and Natural Ventilation in a Hot Arid Climate. In R. González-Lezcano (Ed.), *Advancements in Sustainable Architecture and Energy Efficiency* (pp. 146–168). IGI Global. https://doi.org/10.4018/978-1-7998-7023-4.ch007

Al-Shebeeb, O. A., Rangaswamy, S., Gopalakrishan, B., & Devaru, D. G. (2017). Evaluation and Indexing of Process Plans Based on Electrical Demand and Energy Consumption. *International Journal of Manufacturing, Materials, and Mechanical Engineering*, 7(3), 1–19. doi:10.4018/IJMMME.2017070101

Amuda, M. O., Lawal, T. F., & Akinlabi, E. T. (2017). Research Progress on Rheological Behavior of AA7075 Aluminum Alloy During Hot Deformation. *International Journal of Materials Forming and Machining Processes*, 4(1), 53–96. doi:10.4018/IJMFMP.2017010104

Amuda, M. O., Lawal, T. F., & Mridha, S. (2021). Microstructure and Mechanical Properties of Silicon Carbide-Treated Ferritic Stainless Steel Welds. In L. Burstein (Ed.), *Handbook of Research on Advancements in Manufacturing, Materials, and Mechanical Engineering* (pp. 395–411). IGI Global. https://doi.org/10.4018/978-1-7998-4939-1.ch019

Anikeev, V., Gasem, K. A., & Fan, M. (2021). Application of Supercritical Technologies in Clean Energy Production: A Review. In L. Chen (Ed.), *Handbook of Research on Advancements in Supercritical Fluids Applications for Sustainable Energy Systems* (pp. 792–821). IGI Global. https://doi.org/10.4018/978-1-7998-5796-9.ch022

Arafat, M. Y., Saleem, I., & Devi, T. P. (2022). Drivers of EV Charging Infrastructure Entrepreneurship in India. In M. Alam, R. Pillai, & N. Murugesan (Eds.), *Developing Charging Infrastructure and Technologies for Electric Vehicles* (pp. 208–219). IGI Global. https://doi.org/10.4018/978-1-7998-6858-3.ch011

Araujo, A., & Manninen, H. (2022). Contribution of Project-Based Learning on Social Skills Development: An Industrial Engineer Perspective. In A. Alves & N. van Hattum-Janssen (Eds.), *Training Engineering Students for Modern Technological Advancement* (pp. 119–145). IGI Global. https://doi.org/10.4018/978-1-7998-8816-1.ch006

Armutlu, H. (2018). Intelligent Biomedical Engineering Operations by Cloud Computing Technologies. In U. Kose, G. Guraksin, & O. Deperlioglu (Eds.), *Nature-Inspired Intelligent Techniques for Solving Biomedical Engineering Problems* (pp. 297–317). Hershey, PA: IGI Global. doi:10.4018/978-1-5225-4769-3.ch015

Atik, M., Sadek, M., & Shahrour, I. (2017). Single-Run Adaptive Pushover Procedure for Shear Wall Structures. In V. Plevris, G. Kremmyda, & Y. Fahjan (Eds.), *Performance-Based Seismic Design of Concrete Structures and Infrastructures* (pp. 59–83). Hershey, PA: IGI Global. doi:10.4018/978-1-5225-2089-4.ch003

Attia, H. (2021). Smart Power Microgrid Impact on Sustainable Building. In R. González-Lezcano (Ed.), *Advancements in Sustainable Architecture and Energy Efficiency* (pp. 169–194). IGI Global. https://doi.org/10.4018/978-1-7998-7023-4.ch008

Aydin, A., Akyol, E., Gungor, M., Kaya, A., & Tasdelen, S. (2018). Geophysical Surveys in Engineering Geology Investigations With Field Examples. In N. Ceryan (Ed.), *Handbook of Research on Trends and Digital Advances in Engineering Geology* (pp. 257–280). Hershey, PA: IGI Global. doi:10.4018/978-1-5225-2709-1.ch007

Ayoobkhan, M. U. D., Y., A., J., Easwaran, B., & R., T. (2021). Smart Connected Digital Products and IoT Platform With the Digital Twin. In P. Vasant, G. Weber, & W. Punurai (Ed.), Research Advancements in Smart Technology, Optimization, and Renewable Energy (pp. 330-350). IGI Global. https://doi.org/ doi:10.4018/978-1-7998-3970-5.ch016

Baeza Moyano, D., & González Lezcano, R. A. (2021). The Importance of Light in Our Lives: Towards New Lighting in Schools. In R. González-Lezcano (Ed.), *Advancements in Sustainable Architecture and Energy Efficiency* (pp. 239–256). IGI Global. https://doi.org/10.4018/978-1-7998-7023-4.ch011

Bagdadee, A. H. (2021). A Brief Assessment of the Energy Sector of Bangladesh. *International Journal of Energy Optimization and Engineering*, *10*(1), 36–55. doi:10.4018/IJEOE.2021010103

Baklezos, A. T., & Hadjigeorgiou, N. G. (2021). Magnetic Sensors for Space Applications and Magnetic Cleanliness Considerations. In C. Nikolopoulos (Ed.), *Recent Trends on Electromagnetic Environmental Effects for Aeronautics and Space Applications* (pp. 147–185). IGI Global. https://doi.org/10.4018/978-1-7998-4879-0.ch006

Bas, T. G. (2017). Nutraceutical Industry with the Collaboration of Biotechnology and Nutrigenomics Engineering: The Significance of Intellectual Property in the Entrepreneurship and Scientific Research Ecosystems. In T. Bas & J. Zhao (Eds.), *Comparative Approaches to Biotechnology Development and Use in Developed and Emerging Nations* (pp. 1–17). Hershey, PA: IGI Global. doi:10.4018/978-1-5225-1040-6.ch001

Bazeer Ahamed, B., & Periakaruppan, S. (2021). Taxonomy of Influence Maximization Techniques in Unknown Social Networks. In P. Vasant, G. Weber, & W. Punurai (Eds.), *Research Advancements in Smart Technology, Optimization, and Renewable Energy* (pp. 351-363). IGI Global. https://doi.org/10.4018/978-1-7998-3970-5.ch017

Beale, R., & André, J. (2017). *Design Solutions and Innovations in Temporary Structures*. Hershey, PA: IGI Global. doi:10.4018/978-1-5225-2199-0

Behnam, B. (2017). Simulating Post-Earthquake Fire Loading in Conventional RC Structures. In P. Samui, S. Chakraborty, & D. Kim (Eds.), *Modeling and Simulation Techniques in Structural Engineering* (pp. 425–444). Hershey, PA: IGI Global. doi:10.4018/978-1-5225-0588-4.ch015

Ben Hamida, I., Salah, S. B., Msahli, F., & Mimouni, M. F. (2018). Distribution Network Reconfiguration Using SPEA2 for Power Loss Minimization and Reliability Improvement. *International Journal of Energy Optimization and Engineering, 7*(1), 50–65. doi:10.4018/IJEOE.2018010103

Bentarzi, H. (2021). Fault Tree-Based Root Cause Analysis Used to Study Mal-Operation of a Protective Relay in a Smart Grid. In A. Recioui & H. Bentarzi (Eds.), *Optimizing and Measuring Smart Grid Operation and Control* (pp. 289–308). IGI Global. https://doi.org/10.4018/978-1-7998-4027-5.ch012

Beysens, D. A., Garrabos, Y., & Zappoli, B. (2021). Thermal Effects in Near-Critical Fluids: Piston Effect and Related Phenomena. In L. Chen (Ed.), *Handbook of Research on Advancements in Supercritical Fluids Applications for Sustainable Energy Systems* (pp. 1–31). IGI Global. https://doi.org/10.4018/978-1-7998-5796-9.ch001

Bhaskar, S. V., & Kudal, H. N. (2017). Effect of TiCN and AlCrN Coating on Tribological Behaviour of Plasma-nitrided AISI 4140 Steel. *International Journal of Surface Engineering and Interdisciplinary Materials Science, 5*(2), 1–17. doi:10.4018/IJSEIMS.2017070101

Bhuyan, D. (2018). Designing of a Twin Tube Shock Absorber: A Study in Reverse Engineering. In K. Kumar & J. Davim (Eds.), *Design and Optimization of Mechanical Engineering Products* (pp. 83–104). Hershey, PA: IGI Global. doi:10.4018/978-1-5225-3401-3.ch005

Blumberg, G. (2021). Blockchains for Use in Construction and Engineering Projects. In J. Underwood & M. Shelbourn (Eds.), *Handbook of Research on Driving Transformational Change in the Digital Built Environment* (pp. 179–208). IGI Global. https://doi.org/10.4018/978-1-7998-6600-8.ch008

Bolboaca, A. M. (2021). Considerations Regarding the Use of Fuel Cells in Combined Heat and Power for Stationary Applications. In G. Badea, R. Felseghi, & I. Aşchilean (Eds.), *Hydrogen Fuel Cell Technology for Stationary Applications* (pp. 239–275). IGI Global. https://doi.org/10.4018/978-1-7998-4945-2.ch010

Burstein, L. (2021). Simulation Tool for Cable Design. In L. Burstein (Ed.), *Handbook of Research on Advancements in Manufacturing, Materials, and Mechanical Engineering* (pp. 54–74). IGI Global. https://doi.org/10.4018/978-1-7998-4939-1.ch003

Calderon, F. A., Giolo, E. G., Frau, C. D., Rengel, M. G., Rodriguez, H., Tornello, M., ... Gallucci, R. (2018). Seismic Microzonation and Site Effects Detection Through Microtremors Measures: A Review. In N. Ceryan (Ed.), *Handbook of Research on Trends and Digital Advances in Engineering Geology* (pp. 326–349). Hershey, PA: IGI Global. doi:10.4018/978-1-5225-2709-1.ch009

Ceryan, N., & Can, N. K. (2018). Prediction of The Uniaxial Compressive Strength of Rocks Materials. In N. Ceryan (Ed.), *Handbook of Research on Trends and Digital Advances in Engineering Geology* (pp. 31–96). Hershey, PA: IGI Global. doi:10.4018/978-1-5225-2709-1.ch002

Ceryan, S. (2018). Weathering Indices Used in Evaluation of the Weathering State of Rock Material. In N. Ceryan (Ed.), *Handbook of Research on Trends and Digital Advances in Engineering Geology* (pp. 132–186). Hershey, PA: IGI Global. doi:10.4018/978-1-5225-2709-1.ch004

Chen, H., Padilla, R. V., & Besarati, S. (2017). Supercritical Fluids and Their Applications in Power Generation. In L. Chen & Y. Iwamoto (Eds.), *Advanced Applications of Supercritical Fluids in Energy Systems* (pp. 369–402). Hershey, PA: IGI Global. doi:10.4018/978-1-5225-2047-4.ch012

Chen, H., Padilla, R. V., & Besarati, S. (2021). Supercritical Fluids and Their Applications in Power Generation. In L. Chen (Ed.), *Handbook of Research on Advancements in Supercritical Fluids Applications for Sustainable Energy Systems* (pp. 566–599). IGI Global. https://doi.org/10.4018/978-1-7998-5796-9.ch016

Chen, L. (2017). Principles, Experiments, and Numerical Studies of Supercritical Fluid Natural Circulation System. In L. Chen & Y. Iwamoto (Eds.), *Advanced Applications of Supercritical Fluids in Energy Systems* (pp. 136–187). Hershey, PA: IGI Global. doi:10.4018/978-1-5225-2047-4.ch005

Chen, L. (2021). Principles, Experiments, and Numerical Studies of Supercritical Fluid Natural Circulation System. In L. Chen (Ed.), *Handbook of Research on Advancements in Supercritical Fluids Applications for Sustainable Energy Systems* (pp. 219–269). IGI Global. https://doi.org/10.4018/978-1-7998-5796-9.ch007

Chiba, Y., Marif, Y., Henini, N., & Tlemcani, A. (2021). Modeling of Magnetic Refrigeration Device by Using Artificial Neural Networks Approach. *International Journal of Energy Optimization and Engineering*, *10*(4), 68–76. https://doi.org/10.4018/IJEOE.2021100105

Clementi, F., Di Sciascio, G., Di Sciascio, S., & Lenci, S. (2017). Influence of the Shear-Bending Interaction on the Global Capacity of Reinforced Concrete Frames: A Brief Overview of the New Perspectives. In V. Plevris, G. Kremmyda, & Y. Fahjan (Eds.), *Performance-Based Seismic Design of Concrete Structures and Infrastructures* (pp. 84–111). Hershey, PA: IGI Global. doi:10.4018/978-1-5225-2089-4.ch004

Codinhoto, R., Fialho, B. C., Pinti, L., & Fabricio, M. M. (2021). BIM and IoT for Facilities Management: Understanding Key Maintenance Issues. In J. Underwood & M. Shelbourn (Eds.), *Handbook of Research on Driving Transformational Change in the Digital Built Environment* (pp. 209–231). IGI Global. doi:10.4018/978-1-7998-6600-8.ch009

Cortés-Polo, D., Calle-Cancho, J., Carmona-Murillo, J., & González-Sánchez, J. (2017). Future Trends in Mobile-Fixed Integration for Next Generation Networks: Classification and Analysis. *International Journal of Vehicular Telematics and Infotainment Systems*, *1*(1), 33–53. doi:10.4018/IJVTIS.2017010103

Costa, H. G., Sheremetieff, F. H., & Araújo, E. A. (2022). Influence of Game-Based Methods in Developing Engineering Competences. In A. Alves & N. van Hattum-Janssen (Eds.), *Training Engineering Students for Modern Technological Advancement* (pp. 69–88). IGI Global. https://doi.org/10.4018/978-1-7998-8816-1.ch004

Cui, X., Zeng, S., Li, Z., Zheng, Q., Yu, X., & Han, B. (2018). Advanced Composites for Civil Engineering Infrastructures. In K. Kumar & J. Davim (Eds.), *Composites and Advanced Materials for Industrial Applications* (pp. 212–248). Hershey, PA: IGI Global. doi:10.4018/978-1-5225-5216-1.ch010

Dalgıç, S., & Kuşku, İ. (2018). Geological and Geotechnical Investigations in Tunneling. In N. Ceryan (Ed.), *Handbook of Research on Trends and Digital Advances in Engineering Geology* (pp. 482–529). Hershey, PA: IGI Global. doi:10.4018/978-1-5225-2709-1.ch014

Dang, C., & Hihara, E. (2021). Study on Cooling Heat Transfer of Supercritical Carbon Dioxide Applied to Transcritical Carbon Dioxide Heat Pump. In L. Chen (Ed.), *Handbook of Research on Advancements in Supercritical Fluids Applications for Sustainable Energy Systems* (pp. 451–493). IGI Global. https://doi.org/10.4018/978-1-7998-5796-9.ch013

Daus, Y., Kharchenko, V., & Yudaev, I. (2021). Research of Solar Energy Potential of Photovoltaic Installations on Enclosing Structures of Buildings. *International Journal of Energy Optimization and Engineering*, *10*(4), 18–34. https://doi.org/10.4018/IJEOE.2021100102

Daus, Y., Kharchenko, V., & Yudaev, I. (2021). Optimizing Layout of Distributed Generation Sources of Power Supply System of Agricultural Object. *International Journal of Energy Optimization and Engineering*, *10*(3), 70–84. https://doi.org/10.4018/IJEOE.2021070104

de la Varga, D., Soto, M., Arias, C. A., van Oirschot, D., Kilian, R., Pascual, A., & Álvarez, J. A. (2017). Constructed Wetlands for Industrial Wastewater Treatment and Removal of Nutrients. In Á. Val del Río, J. Campos Gómez, & A. Mosquera Corral (Eds.), *Technologies for the Treatment and Recovery of Nutrients from Industrial Wastewater* (pp. 202–230). Hershey, PA: IGI Global. doi:10.4018/978-1-5225-1037-6.ch008

Deb, S., Ammar, E. A., AlRajhi, H., Alsaidan, I., & Shariff, S. M. (2022). V2G Pilot Projects: Review and Lessons Learnt. In M. Alam, R. Pillai, & N. Murugesan (Eds.), *Developing Charging Infrastructure and Technologies for Electric Vehicles* (pp. 252–267). IGI Global. https://doi.org/10.4018/978-1-7998-6858-3.ch014

Dekhandji, F. Z., & Rais, M. C. (2021). A Comparative Study of Power Quality Monitoring Using Various Techniques. In A. Recioui & H. Bentarzi (Eds.), *Optimizing and Measuring Smart Grid Operation and Control* (pp. 259–288). IGI Global. https://doi.org/10.4018/978-1-7998-4027-5.ch011

Deperlioglu, O. (2018). Intelligent Techniques Inspired by Nature and Used in Biomedical Engineering. In U. Kose, G. Guraksin, & O. Deperlioglu (Eds.), *Nature-Inspired Intelligent Techniques for Solving Biomedical Engineering Problems* (pp. 51–77). Hershey, PA: IGI Global. doi:10.4018/978-1-5225-4769-3.ch003

Dhurpate, P. R., & Tang, H. (2021). Quantitative Analysis of the Impact of Inter-Line Conveyor Capacity for Throughput of Manufacturing Systems. *International Journal of Manufacturing, Materials, and Mechanical Engineering*, *11*(1), 1–17. https://doi.org/10.4018/IJMMME.2021010101

Dinkar, S., & Deep, K. (2021). A Survey of Recent Variants and Applications of Antlion Optimizer. *International Journal of Energy Optimization and Engineering*, *10*(2), 48–73. doi:10.4018/IJEOE.2021040103

Dixit, A. (2018). Application of Silica-Gel-Reinforced Aluminium Composite on the Piston of Internal Combustion Engine: Comparative Study of Silica-Gel-Reinforced Aluminium Composite Piston With Aluminium Alloy Piston. In K. Kumar & J. Davim (Eds.), *Composites and Advanced Materials for Industrial Applications* (pp. 63–98). Hershey, PA: IGI Global. doi:10.4018/978-1-5225-5216-1.ch004

Drabecki, M. P., & Kułak, K. B. (2021). Global Pandemics on European Electrical Energy Markets: Lessons Learned From the COVID-19 Outbreak. *International Journal of Energy Optimization and Engineering*, *10*(3), 24–46. https://doi.org/10.4018/IJEOE.2021070102

Dutta, M. M. (2021). Nanomaterials for Food and Agriculture. In M. Bhat, I. Wani, & S. Ashraf (Eds.), *Applications of Nanomaterials in Agriculture, Food Science, and Medicine* (pp. 75–97). IGI Global. doi:10.4018/978-1-7998-5563-7.ch004

Dutta, M. M., & Goswami, M. (2021). Coating Materials: Nano-Materials. In S. Roy & G. Bose (Eds.), *Advanced Surface Coating Techniques for Modern Industrial Applications* (pp. 1–30). IGI Global. doi:10.4018/978-1-7998-4870-7.ch001

Elsayed, A. M., Dakkama, H. J., Mahmoud, S., Al-Dadah, R., & Kaialy, W. (2017). Sustainable Cooling Research Using Activated Carbon Adsorbents and Their Environmental Impact. In T. Kobayashi (Ed.), *Applied Environmental Materials Science for Sustainability* (pp. 186–221). Hershey, PA: IGI Global. doi:10.4018/978-1-5225-1971-3.ch009

Ercanoglu, M., & Sonmez, H. (2018). General Trends and New Perspectives on Landslide Mapping and Assessment Methods. In N. Ceryan (Ed.), *Handbook of Research on Trends and Digital Advances in Engineering Geology* (pp. 350–379). Hershey, PA: IGI Global. doi:10.4018/978-1-5225-2709-1.ch010

Faroz, S. A., Pujari, N. N., Rastogi, R., & Ghosh, S. (2017). Risk Analysis of Structural Engineering Systems Using Bayesian Inference. In P. Samui, S. Chakraborty, & D. Kim (Eds.), *Modeling and Simulation Techniques in Structural Engineering* (pp. 390–424). Hershey, PA: IGI Global. doi:10.4018/978-1-5225-0588-4.ch014

Fekik, A., Hamida, M. L., Denoun, H., Azar, A. T., Kamal, N. A., Vaidyanathan, S., Bousbaine, A., & Benamrouche, N. (2022). Multilevel Inverter for Hybrid Fuel Cell/PV Energy Conversion System. In A. Fekik & N. Benamrouche (Eds.), *Modeling and Control of Static Converters for Hybrid Storage Systems* (pp. 233–270). IGI Global. https://doi.org/10.4018/978-1-7998-7447-8.ch009

Fekik, A., Hamida, M. L., Houassine, H., Azar, A. T., Kamal, N. A., Denoun, H., Vaidyanathan, S., & Sambas, A. (2022). Power Quality Improvement for Grid-Connected Photovoltaic Panels Using Direct Power Control. In A. Fekik & N. Benamrouche (Eds.), *Modeling and Control of Static Converters for Hybrid Storage Systems* (pp. 107–142). IGI Global. https://doi.org/10.4018/978-1-7998-7447-8.ch005

Fernando, P. R., Hamigah, T., Disne, S., Wickramasingha, G. G., & Sutharshan, A. (2018). The Evaluation of Engineering Properties of Low Cost Concrete Blocks by Partial Doping of Sand with Sawdust: Low Cost Sawdust Concrete Block. *International Journal of Strategic Engineering*, *1*(2), 26–42. doi:10.4018/IJoSE.2018070103

Ferro, G., Minciardi, R., Parodi, L., & Robba, M. (2022). Optimal Charging Management of Microgrid-Integrated Electric Vehicles. In M. Alam, R. Pillai, & N. Murugesan (Eds.), *Developing Charging Infrastructure and Technologies for Electric Vehicles* (pp. 133–155). IGI Global. https://doi.org/10.4018/978-1-7998-6858-3.ch007

Flumerfelt, S., & Green, C. (2022). Graduate Lean Leadership Education: A Case Study of a Program. In A. Alves & N. van Hattum-Janssen (Eds.), *Training Engineering Students for Modern Technological Advancement* (pp. 202–224). IGI Global. https://doi.org/10.4018/978-1-7998-8816-1.ch010

Galli, B. J. (2021). Implications of Economic Decision Making to the Project Manager. *International Journal of Strategic Engineering*, *4*(1), 19–32. https://doi.org/10.4018/IJoSE.2021010102

Gento, A. M., Pimentel, C., & Pascual, J. A. (2022). Teaching Circular Economy and Lean Management in a Learning Factory. In A. Alves & N. van Hattum-Janssen (Eds.), *Training Engineering Students for Modern Technological Advancement* (pp. 183–201). IGI Global. https://doi.org/10.4018/978-1-7998-8816-1.ch009

Ghosh, S., Mitra, S., Ghosh, S., & Chakraborty, S. (2017). Seismic Reliability Analysis in the Framework of Metamodelling Based Monte Carlo Simulation. In P. Samui, S. Chakraborty, & D. Kim (Eds.), *Modeling and Simulation Techniques in Structural Engineering* (pp. 192–208). Hershey, PA: IGI Global. doi:10.4018/978-1-5225-0588-4.ch006

Gil, M., & Otero, B. (2017). Learning Engineering Skills through Creativity and Collaboration: A Game-Based Proposal. In R. Alexandre Peixoto de Queirós & M. Pinto (Eds.), *Gamification-Based E-Learning Strategies for Computer Programming Education* (pp. 14–29). Hershey, PA: IGI Global. doi:10.4018/978-1-5225-1034-5.ch002

Gill, J., Ayre, M., & Mills, J. (2017). Revisioning the Engineering Profession: How to Make It Happen! In M. Gray & K. Thomas (Eds.), *Strategies for Increasing Diversity in Engineering Majors and Careers* (pp. 156–175). Hershey, PA: IGI Global. doi:10.4018/978-1-5225-2212-6.ch008

Godzhaev, Z., Senkevich, S., Kuzmin, V., & Melikov, I. (2021). Use of the Neural Network Controller of Sprung Mass to Reduce Vibrations From Road Irregularities. In P. Vasant, G. Weber, & W. Punurai (Ed.), *Research Advancements in Smart Technology, Optimization, and Renewable Energy* (pp. 69-87). IGI Global. https://doi.org/10.4018/978-1-7998-3970-5.ch005

Gomes de Gusmão, C. M. (2022). Digital Competencies and Transformation in Higher Education: Upskilling With Extension Actions. In A. Alves & N. van Hattum-Janssen (Eds.), *Training Engineering Students for Modern Technological Advancement* (pp. 313–328). IGI Global. https://doi.org/10.4018/978-1-7998-8816-1.ch015A

Goyal, N., Ram, M., & Kumar, P. (2017). Welding Process under Fault Coverage Approach for Reliability and MTTF. In M. Ram & J. Davim (Eds.), *Mathematical Concepts and Applications in Mechanical Engineering and Mechatronics* (pp. 222–245). Hershey, PA: IGI Global. doi:10.4018/978-1-5225-1639-2.ch011

Gray, M., & Lundy, C. (2017). Engineering Study Abroad: High Impact Strategy for Increasing Access. In M. Gray & K. Thomas (Eds.), *Strategies for Increasing Diversity in Engineering Majors and Careers* (pp. 42–59). Hershey, PA: IGI Global. doi:10.4018/978-1-5225-2212-6.ch003

Güler, O., & Varol, T. (2021). Fabrication of Functionally Graded Metal and Ceramic Powders Synthesized by Electroless Deposition. In S. Roy & G. Bose (Eds.), *Advanced Surface Coating Techniques for Modern Industrial Applications* (pp. 150–187). IGI Global. https://doi.org/10.4018/978-1-7998-4870-7.ch007

Guraksin, G. E. (2018). Internet of Things and Nature-Inspired Intelligent Techniques for the Future of Biomedical Engineering. In U. Kose, G. Guraksin, & O. Deperlioglu (Eds.), *Nature-Inspired Intelligent Techniques for Solving Biomedical Engineering Problems* (pp. 263–282). Hershey, PA: IGI Global. doi:10.4018/978-1-5225-4769-3.ch013

Hamida, M. L., Fekik, A., Denoun, H., Ardjal, A., & Bokhtache, A. A. (2022). Flying Capacitor Inverter Integration in a Renewable Energy System. In A. Fekik & N. Benamrouche (Eds.), *Modeling and Control of Static Converters for Hybrid Storage Systems* (pp. 287–306). IGI Global. https://doi.org/10.4018/978-1-7998-7447-8.ch011

Hasegawa, N., & Takahashi, Y. (2021). Control of Soap Bubble Ejection Robot Using Facial Expressions. *International Journal of Manufacturing, Materials, and Mechanical Engineering, 11*(2), 1–16. https://doi.org/10.4018/IJMMME.2021040101

Hejazi, T., & Akbari, L. (2017). A Multiresponse Optimization Model for Statistical Design of Processes with Discrete Variables. In M. Ram & J. Davim (Eds.), *Mathematical Concepts and Applications in Mechanical Engineering and Mechatronics* (pp. 17–37). Hershey, PA: IGI Global. doi:10.4018/978-1-5225-1639-2.ch002

Hejazi, T., & Hejazi, A. (2017). Monte Carlo Simulation for Reliability-Based Design of Automotive Complex Subsystems. In M. Ram & J. Davim (Eds.), *Mathematical Concepts and Applications in Mechanical Engineering and Mechatronics* (pp. 177–200). Hershey, PA: IGI Global. doi:10.4018/978-1-5225-1639-2.ch009

Hejazi, T., & Poursabbagh, H. (2017). Reliability Analysis of Engineering Systems: An Accelerated Life Testing for Boiler Tubes. In M. Ram & J. Davim (Eds.), *Mathematical Concepts and Applications in Mechanical Engineering and Mechatronics* (pp. 154–176). Hershey, PA: IGI Global. doi:10.4018/978-1-5225-1639-2.ch008

Henao, J., Poblano-Salas, C. A., Vargas, F., Giraldo-Betancur, A. L., Corona-Castuera, J., & Sotelo-Mazón, O. (2021). Principles and Applications of Thermal Spray Coatings. In S. Roy & G. Bose (Eds.), *Advanced Surface Coating Techniques for Modern Industrial Applications* (pp. 31–70). IGI Global. https://doi.org/10.4018/978-1-7998-4870-7.ch002

Henao, J., & Sotelo, O. (2018). Surface Engineering at High Temperature: Thermal Cycling and Corrosion Resistance. In A. Pakseresht (Ed.), *Production, Properties, and Applications of High Temperature Coatings* (pp. 131–159). Hershey, PA: IGI Global. doi:10.4018/978-1-5225-4194-3.ch006

Hrnčič, M. K., Cör, D., & Knez, Ž. (2021). Supercritical Fluids as a Tool for Green Energy and Chemicals. In L. Chen (Ed.), *Handbook of Research on Advancements in Supercritical Fluids Applications for Sustainable Energy Systems* (pp. 761–791). IGI Global. doi:10.4018/978-1-7998-5796-9.ch021

Ibrahim, O., Erdem, S., & Gurbuz, E. (2021). Studying Physical and Chemical Properties of Graphene Oxide and Reduced Graphene Oxide and Their Applications in Sustainable Building Materials. In R. González-Lezcano (Ed.), *Advancements in Sustainable Architecture and Energy Efficiency* (pp. 221–238). IGI Global. https://doi.org/10.4018/978-1-7998-7023-4.ch010

Ihianle, I. K., Islam, S., Naeem, U., & Ebenuwa, S. H. (2021). Exploiting Patterns of Object Use for Human Activity Recognition. In A. Nwajana & I. Ihianle (Eds.), *Handbook of Research on 5G Networks and Advancements in Computing, Electronics, and Electrical Engineering* (pp. 382–401). IGI Global. https://doi.org/10.4018/978-1-7998-6992-4.ch015

Ijemaru, G. K., Ngharamike, E. T., Oleka, E. U., & Nwajana, A. O. (2021). An Energy-Efficient Model for Opportunistic Data Collection in IoV-Enabled SC Waste Management. In A. Nwajana & I. Ihianle (Eds.), *Handbook of Research on 5G Networks and Advancements in Computing, Electronics, and Electrical Engineering* (pp. 1–19). IGI Global. https://doi.org/10.4018/978-1-7998-6992-4.ch001

Ilori, O. O., Adetan, D. A., & Umoru, L. E. (2017). Effect of Cutting Parameters on the Surface Residual Stress of Face-Milled Pearlitic Ductile Iron. *International Journal of Materials Forming and Machining Processes*, *4*(1), 38–52. doi:10.4018/IJMFMP.2017010103

Imam, M. H., Tasadduq, I. A., Ahmad, A., Aldosari, F., & Khan, H. (2017). Automated Generation of Course Improvement Plans Using Expert System. *International Journal of Quality Assurance in Engineering and Technology Education*, *6*(1), 1–12. doi:10.4018/IJQAETE.2017010101

Injeti, S. K., & Kumar, T. V. (2018). A WDO Framework for Optimal Deployment of DGs and DSCs in a Radial Distribution System Under Daily Load Pattern to Improve Techno-Economic Benefits. *International Journal of Energy Optimization and Engineering*, *7*(2), 1–38. doi:10.4018/IJEOE.2018040101

Ishii, N., Anami, K., & Knisely, C. W. (2018). *Dynamic Stability of Hydraulic Gates and Engineering for Flood Prevention*. Hershey, PA: IGI Global. doi:10.4018/978-1-5225-3079-4

Iwamoto, Y., & Yamaguchi, H. (2021). Application of Supercritical Carbon Dioxide for Solar Water Heater. In L. Chen (Ed.), *Handbook of Research on Advancements in Supercritical Fluids Applications for Sustainable Energy Systems* (pp. 370–387). IGI Global. https://doi.org/10.4018/978-1-7998-5796-9.ch010

Jayapalan, S. (2018). A Review of Chemical Treatments on Natural Fibers-Based Hybrid Composites for Engineering Applications. In K. Kumar & J. Davim (Eds.), *Composites and Advanced Materials for Industrial Applications* (pp. 16–37). Hershey, PA: IGI Global. doi:10.4018/978-1-5225-5216-1.ch002

Kapetanakis, T. N., Vardiambasis, I. O., Ioannidou, M. P., & Konstantaras, A. I. (2021). Modeling Antenna Radiation Using Artificial Intelligence Techniques: The Case of a Circular Loop Antenna. In C. Nikolopoulos (Ed.), *Recent Trends on Electromagnetic Environmental Effects for Aeronautics and Space Applications* (pp. 186–225). IGI Global. https://doi.org/10.4018/978-1-7998-4879-0.ch007

Karkalos, N. E., Markopoulos, A. P., & Dossis, M. F. (2017). Optimal Model Parameters of Inverse Kinematics Solution of a 3R Robotic Manipulator Using ANN Models. *International Journal of Manufacturing, Materials, and Mechanical Engineering*, 7(3), 20–40. doi:10.4018/IJMMME.2017070102

Kelly, M., Costello, M., Nicholson, G., & O'Connor, J. (2021). The Evolving Integration of BIM Into Built Environment Programmes in a Higher Education Institute. In J. Underwood & M. Shelbourn (Eds.), *Handbook of Research on Driving Transformational Change in the Digital Built Environment* (pp. 294–326). IGI Global. https://doi.org/10.4018/978-1-7998-6600-8.ch012

Kesimal, A., Karaman, K., Cihangir, F., & Ercikdi, B. (2018). Excavatability Assessment of Rock Masses for Geotechnical Studies. In N. Ceryan (Ed.), *Handbook of Research on Trends and Digital Advances in Engineering Geology* (pp. 231–256). Hershey, PA: IGI Global. doi:10.4018/978-1-5225-2709-1.ch006

Knoflacher, H. (2017). The Role of Engineers and Their Tools in the Transport Sector after Paradigm Change: From Assumptions and Extrapolations to Science. In H. Knoflacher & E. Ocalir-Akunal (Eds.), *Engineering Tools and Solutions for Sustainable Transportation Planning* (pp. 1–29). Hershey, PA: IGI Global. doi:10.4018/978-1-5225-2116-7.ch001

Kose, U. (2018). Towards an Intelligent Biomedical Engineering With Nature-Inspired Artificial Intelligence Techniques. In U. Kose, G. Guraksin, & O. Deperlioglu (Eds.), *Nature-Inspired Intelligent Techniques for Solving Biomedical Engineering Problems* (pp. 1–26). Hershey, PA: IGI Global. doi:10.4018/978-1-5225-4769-3.ch001

Kostić, S. (2018). A Review on Enhanced Stability Analyses of Soil Slopes Using Statistical Design. In N. Ceryan (Ed.), *Handbook of Research on Trends and Digital Advances in Engineering Geology* (pp. 446–481). Hershey, PA: IGI Global. doi:10.4018/978-1-5225-2709-1.ch013

Kumar, A., Patil, P. P., & Prajapati, Y. K. (2018). *Advanced Numerical Simulations in Mechanical Engineering*. Hershey, PA: IGI Global. doi:10.4018/978-1-5225-3722-9

Kumar, G. R., Rajyalakshmi, G., & Manupati, V. K. (2017). Surface Micro Patterning of Aluminium Reinforced Composite through Laser Peening. *International Journal of Manufacturing, Materials, and Mechanical Engineering, 7*(4), 15–27. doi:10.4018/IJMMME.2017100102

Kumar, N., Basu, D. N., & Chen, L. (2021). Effect of Flow Acceleration and Buoyancy on Thermalhydraulics of sCO2 in Mini/Micro-Channel. In L. Chen (Ed.), *Handbook of Research on Advancements in Supercritical Fluids Applications for Sustainable Energy Systems* (pp. 161–182). IGI Global. doi:10.4018/978-1-7998-5796-9.ch005

Kumari, N., & Kumar, K. (2018). Fabrication of Orthotic Calipers With Epoxy-Based Green Composite. In K. Kumar & J. Davim (Eds.), *Composites and Advanced Materials for Industrial Applications* (pp. 157–176). Hershey, PA: IGI Global. doi:10.4018/978-1-5225-5216-1.ch008

Kuppusamy, R. R. (2018). Development of Aerospace Composite Structures Through Vacuum-Enhanced Resin Transfer Moulding Technology (VERTMTy): Vacuum-Enhanced Resin Transfer Moulding. In K. Kumar & J. Davim (Eds.), *Composites and Advanced Materials for Industrial Applications* (pp. 99–111). Hershey, PA: IGI Global. doi:10.4018/978-1-5225-5216-1.ch005

Kurganov, V. A., Zeigarnik, Y. A., & Maslakova, I. V. (2021). Normal and Deteriorated Heat Transfer Under Heating Turbulent Supercritical Pressure Coolants Flows in Round Tubes. In L. Chen (Ed.), *Handbook of Research on Advancements in Supercritical Fluids Applications for Sustainable Energy Systems* (pp. 494–532). IGI Global. https://doi.org/10.4018/978-1-7998-5796-9.ch014

Li, H., & Zhang, Y. (2021). Heat Transfer and Fluid Flow Modeling for Supercritical Fluids in Advanced Energy Systems. In L. Chen (Ed.), *Handbook of Research on Advancements in Supercritical Fluids Applications for Sustainable Energy Systems* (pp. 388–422). IGI Global. https://doi.org/10.4018/978-1-7998-5796-9.ch011

Loy, J., Howell, S., & Cooper, R. (2017). Engineering Teams: Supporting Diversity in Engineering Education. In M. Gray & K. Thomas (Eds.), *Strategies for Increasing Diversity in Engineering Majors and Careers* (pp. 106–129). Hershey, PA: IGI Global. doi:10.4018/978-1-5225-2212-6.ch006

Macher, G., Armengaud, E., Kreiner, C., Brenner, E., Schmittner, C., Ma, Z., ... Krammer, M. (2018). Integration of Security in the Development Lifecycle of Dependable Automotive CPS. In N. Druml, A. Genser, A. Krieg, M. Menghin, & A. Hoeller (Eds.), *Solutions for Cyber-Physical Systems Ubiquity* (pp. 383–423). Hershey, PA: IGI Global. doi:10.4018/978-1-5225-2845-6.ch015

Madhu, M. N., Singh, J. G., Mohan, V., & Ongsakul, W. (2021). Transmission Risk Optimization in Interconnected Systems: Risk-Adjusted Available Transfer Capability. In P. Vasant, G. Weber, & W. Punurai (Ed.), *Research Advancements in Smart Technology, Optimization, and Renewable Energy* (pp. 183-199). IGI Global. https://doi.org/10.4018/978-1-7998-3970-5.ch010

Mahendramani, G., & Lakshmana Swamy, N. (2018). Effect of Weld Groove Area on Distortion of Butt Welded Joints in Submerged Arc Welding. *International Journal of Manufacturing, Materials, and Mechanical Engineering*, 8(2), 33–44. doi:10.4018/IJMMME.2018040103

Makropoulos, G., Koumaras, H., Setaki, F., Filis, K., Lutz, T., Montowtt, P., Tomaszewski, L., Dybiec, P., & Järvet, T. (2021). 5G and Unmanned Aerial Vehicles (UAVs) Use Cases: Analysis of the Ecosystem, Architecture, and Applications. In A. Nwajana & I. Ihianle (Eds.), *Handbook of Research on 5G Networks and Advancements in Computing, Electronics, and Electrical Engineering* (pp. 36–69). IGI Global. https://doi.org/10.4018/978-1-7998-6992-4.ch003

Meric, E. M., Erdem, S., & Gurbuz, E. (2021). Application of Phase Change Materials in Construction Materials for Thermal Energy Storage Systems in Buildings. In R. González-Lezcano (Ed.), *Advancements in Sustainable Architecture and Energy Efficiency* (pp. 1–20). IGI Global. https://doi.org/10.4018/978-1-7998-7023-4.ch001

Mihret, E. T., & Yitayih, K. A. (2021). Operation of VANET Communications: The Convergence of UAV System With LTE/4G and WAVE Technologies. *International Journal of Smart Vehicles and Smart Transportation*, 4(1), 29–51. https://doi.org/10.4018/IJSVST.2021010103

Mir, M. A., Bhat, B. A., Sheikh, B. A., Rather, G. A., Mehraj, S., & Mir, W. R. (2021). Nanomedicine in Human Health Therapeutics and Drug Delivery: Nanobiotechnology and Nanobiomedicine. In M. Bhat, I. Wani, & S. Ashraf (Eds.), *Applications of Nanomaterials in Agriculture, Food Science, and Medicine* (pp. 229–251). IGI Global. doi:10.4018/978-1-7998-5563-7.ch013

Mohammadzadeh, S., & Kim, Y. (2017). Nonlinear System Identification of Smart Buildings. In P. Samui, S. Chakraborty, & D. Kim (Eds.), *Modeling and Simulation Techniques in Structural Engineering* (pp. 328–347). Hershey, PA: IGI Global. doi:10.4018/978-1-5225-0588-4.ch011

Molina, G. J., Aktaruzzaman, F., Soloiu, V., & Rahman, M. (2017). Design and Testing of a Jet-Impingement Instrument to Study Surface-Modification Effects by Nanofluids. *International Journal of Surface Engineering and Interdisciplinary Materials Science*, 5(2), 43–61. doi:10.4018/IJSEIMS.2017070104

Moreno-Rangel, A., & Carrillo, G. (2021). Energy-Efficient Homes: A Heaven for Respiratory Illnesses. In R. González-Lezcano (Ed.), *Advancements in Sustainable Architecture and Energy Efficiency* (pp. 49–71). IGI Global. https://doi.org/10.4018/978-1-7998-7023-4.ch003

Msomi, V., & Jantjies, B. T. (2021). Correlative Analysis Between Tensile Properties and Tool Rotational Speeds of Friction Stir Welded Similar Aluminium Alloy Joints. *International Journal of Surface Engineering and Interdisciplinary Materials Science*, 9(2), 58–78. https://doi.org/10.4018/IJSEIMS.2021070104

Muigai, M. N., Mwema, F. M., Akinlabi, E. T., & Obiko, J. O. (2021). Surface Engineering of Materials Through Weld-Based Technologies: An Overview. In S. Roy & G. Bose (Eds.), *Advanced Surface Coating Techniques for Modern Industrial Applications* (pp. 247–260). IGI Global. doi:10.4018/978-1-7998-4870-7.ch011

Mukherjee, A., Saeed, R. A., Dutta, S., & Naskar, M. K. (2017). Fault Tracking Framework for Software-Defined Networking (SDN). In C. Singhal & S. De (Eds.), *Resource Allocation in Next-Generation Broadband Wireless Access Networks* (pp. 247–272). Hershey, PA: IGI Global. doi:10.4018/978-1-5225-2023-8.ch011

Mukhopadhyay, A., Barman, T. K., & Sahoo, P. (2018). Electroless Nickel Coatings for High Temperature Applications. In K. Kumar & J. Davim (Eds.), *Composites and Advanced Materials for Industrial Applications* (pp. 297–331). Hershey, PA: IGI Global. doi:10.4018/978-1-5225-5216-1.ch013

Mwema, F. M., & Wambua, J. M. (2022). Machining of Poly Methyl Methacrylate (PMMA) and Other Olymeric Materials: A Review. In K. Kumar, B. Babu, & J. Davim (Eds.), *Handbook of Research on Advancements in the Processing, Characterization, and Application of Lightweight Materials* (pp. 363–379). IGI Global. https://doi.org/10.4018/978-1-7998-7864-3.ch016

Mykhailyshyn, R., Savkiv, V., Boyko, I., Prada, E., & Virgala, I. (2021). Substantiation of Parameters of Friction Elements of Bernoulli Grippers With a Cylindrical Nozzle. *International Journal of Manufacturing, Materials, and Mechanical Engineering*, *11*(2), 17–39. https://doi.org/10.4018/IJMMME.2021040102

Náprstek, J., & Fischer, C. (2017). Dynamic Stability and Post-Critical Processes of Slender Auto-Parametric Systems. In V. Plevris, G. Kremmyda, & Y. Fahjan (Eds.), *Performance-Based Seismic Design of Concrete Structures and Infrastructures* (pp. 128–171). Hershey, PA: IGI Global. doi:10.4018/978-1-5225-2089-4.ch006

Nautiyal, L., Shivach, P., & Ram, M. (2018). Optimal Designs by Means of Genetic Algorithms. In M. Ram & J. Davim (Eds.), *Soft Computing Techniques and Applications in Mechanical Engineering* (pp. 151–161). Hershey, PA: IGI Global. doi:10.4018/978-1-5225-3035-0.ch007

Nazir, R. (2017). Advanced Nanomaterials for Water Engineering and Treatment: Nano-Metal Oxides and Their Nanocomposites. In T. Saleh (Ed.), *Advanced Nanomaterials for Water Engineering, Treatment, and Hydraulics* (pp. 84–126). Hershey, PA: IGI Global. doi:10.4018/978-1-5225-2136-5.ch005

Nikolopoulos, C. D. (2021). Recent Advances on Measuring and Modeling ELF-Radiated Emissions for Space Applications. In C. Nikolopoulos (Ed.), *Recent Trends on Electromagnetic Environmental Effects for Aeronautics and Space Applications* (pp. 1–38). IGI Global. https://doi.org/10.4018/978-1-7998-4879-0.ch001

Nogueira, A. F., Ribeiro, J. C., Fernández de Vega, F., & Zenha-Rela, M. A. (2018). Evolutionary Approaches to Test Data Generation for Object-Oriented Software: Overview of Techniques and Tools. In M. Khosrow-Pour, D.B.A. (Ed.), Incorporating Nature-Inspired Paradigms in Computational Applications (pp. 162-194). Hershey, PA: IGI Global. https://doi.org/ doi:10.4018/978-1-5225-5020-4.ch006

Nwajana, A. O., Obi, E. R., Ijemaru, G. K., Oleka, E. U., & Anthony, D. C. (2021). Fundamentals of RF/Microwave Bandpass Filter Design. In A. Nwajana & I. Ihianle (Eds.), *Handbook of Research on 5G Networks and Advancements in Computing, Electronics, and Electrical Engineering* (pp. 149–164). IGI Global. https://doi.org/10.4018/978-1-7998-6992-4.ch005

Ogbodo, E. A. (2021). Comparative Study of Transmission Line Junction vs. Asynchronously Coupled Junction Diplexers. In A. Nwajana & I. Ihianle (Eds.), *Handbook of Research on 5G Networks and Advancements in Computing, Electronics, and Electrical Engineering* (pp. 326–336). IGI Global. https://doi.org/10.4018/978-1-7998-6992-4.ch013

Orosa, J. A., Vergara, D., Fraguela, F., & Masdías-Bonome, A. (2021). Statistical Understanding and Optimization of Building Energy Consumption and Climate Change Consequences. In R. González-Lezcano (Ed.), *Advancements in Sustainable Architecture and Energy Efficiency* (pp. 195–220). IGI Global. https://doi.org/10.4018/978-1-7998-7023-4.ch009

Osho, M. B. (2018). Industrial Enzyme Technology: Potential Applications. In S. Bharati & P. Chaurasia (Eds.), *Research Advancements in Pharmaceutical, Nutritional, and Industrial Enzymology* (pp. 375–394). Hershey, PA: IGI Global. doi:10.4018/978-1-5225-5237-6.ch017

Ouadi, A., & Zitouni, A. (2021). Phasor Measurement Improvement Using Digital Filter in a Smart Grid. In A. Recioui & H. Bentarzi (Eds.), *Optimizing and Measuring Smart Grid Operation and Control* (pp. 100–117). IGI Global. https://doi.org/10.4018/978-1-7998-4027-5.ch005

Padmaja, P., & Marutheswar, G. (2017). Certain Investigation on Secured Data Transmission in Wireless Sensor Networks. *International Journal of Mobile Computing and Multimedia Communications*, *8*(1), 48–61. doi:10.4018/IJMCMC.2017010104

Palmer, S., & Hall, W. (2017). An Evaluation of Group Work in First-Year Engineering Design Education. In R. Tucker (Ed.), *Collaboration and Student Engagement in Design Education* (pp. 145–168). Hershey, PA: IGI Global. doi:10.4018/978-1-5225-0726-0.ch007

Panchenko, V. (2021). Prospects for Energy Supply of the Arctic Zone Objects of Russia Using Frost-Resistant Solar Modules. In P. Vasant, G. Weber, & W. Punurai (Eds.), *Research Advancements in Smart Technology, Optimization, and Renewable Energy* (pp. 149-169). IGI Global. https://doi.org/10.4018/978-1-7998-3970-5.ch008

Panchenko, V. (2021). Photovoltaic Thermal Module With Paraboloid Type Solar Concentrators. *International Journal of Energy Optimization and Engineering*, *10*(2), 1–23. https://doi.org/10.4018/IJEOE.2021040101

Pandey, K., & Datta, S. (2021). Dry Machining of Inconel 825 Superalloys: Performance of Tool Inserts (Carbide, Cermet, and SiAlON). *International Journal of Manufacturing, Materials, and Mechanical Engineering*, *11*(4), 26–39. doi:10.4018/IJMMME.2021100102

Panneer, R. (2017). Effect of Composition of Fibers on Properties of Hybrid Composites. *International Journal of Manufacturing, Materials, and Mechanical Engineering*, *7*(4), 28–43. doi:10.4018/IJMMME.2017100103

Pany, C. (2021). Estimation of Correct Long-Seam Mismatch Using FEA to Compare the Measured Strain in a Non-Destructive Testing of a Pressurant Tank: A Reverse Problem. *International Journal of Smart Vehicles and Smart Transportation*, 4(1), 16–28. doi:10.4018/IJSVST.2021010102

Paul, S., & Roy, P. (2018). Optimal Design of Power System Stabilizer Using a Novel Evolutionary Algorithm. *International Journal of Energy Optimization and Engineering*, 7(3), 24–46. doi:10.4018/IJEOE.2018070102

Paul, S., & Roy, P. K. (2021). Oppositional Differential Search Algorithm for the Optimal Tuning of Both Single Input and Dual Input Power System Stabilizer. In P. Vasant, G. Weber, & W. Punurai (Eds.), *Research Advancements in Smart Technology, Optimization, and Renewable Energy* (pp. 256-282). IGI Global. https://doi.org/10.4018/978-1-7998-3970-5.ch013

Pavaloiu, A. (2018). Artificial Intelligence Ethics in Biomedical-Engineering-Oriented Problems. In U. Kose, G. Guraksin, & O. Deperlioglu (Eds.), *Nature-Inspired Intelligent Techniques for Solving Biomedical Engineering Problems* (pp. 219–231). Hershey, PA: IGI Global. doi:10.4018/978-1-5225-4769-3.ch010

Pioro, I., Mahdi, M., & Popov, R. (2017). Application of Supercritical Pressures in Power Engineering. In L. Chen & Y. Iwamoto (Eds.), *Advanced Applications of Supercritical Fluids in Energy Systems* (pp. 404–457). Hershey, PA: IGI Global. doi:10.4018/978-1-5225-2047-4.ch013

Plaksina, T., & Gildin, E. (2017). Rigorous Integrated Evolutionary Workflow for Optimal Exploitation of Unconventional Gas Assets. *International Journal of Energy Optimization and Engineering*, 6(1), 101–122. doi:10.4018/IJEOE.2017010106

Popat, J., Kakadiya, H., Tak, L., Singh, N. K., Majeed, M. A., & Mahajan, V. (2021). Reliability of Smart Grid Including Cyber Impact: A Case Study. In R. Singh, A. Singh, A. Dwivedi, & P. Nagabhushan (Eds.), *Computational Methodologies for Electrical and Electronics Engineers* (pp. 163–174). IGI Global. https://doi.org/10.4018/978-1-7998-3327-7.ch013

Quiza, R., La Fé-Perdomo, I., Rivas, M., & Ramtahalsing, V. (2021). Triple Bottom Line-Focused Optimization of Oblique Turning Processes Based on Hybrid Modeling: A Study Case on AISI 1045 Steel Turning. In L. Burstein (Ed.), *Handbook of Research on Advancements in Manufacturing, Materials, and Mechanical Engineering* (pp. 215–241). IGI Global. https://doi.org/10.4018/978-1-7998-4939-1.ch010

Rahmani, M. K. (2022). Blockchain Technology: Principles and Algorithms. In S. Khan, M. Syed, R. Hammad, & A. Bushager (Eds.), *Blockchain Technology and Computational Excellence for Society 5.0* (pp. 16–27). IGI Global. https://doi.org/10.4018/978-1-7998-8382-1.ch002

Ramdani, N., & Azibi, M. (2018). Polymer Composite Materials for Microelectronics Packaging Applications: Composites for Microelectronics Packaging. In K. Kumar & J. Davim (Eds.), *Composites and Advanced Materials for Industrial Applications* (pp. 177–211). Hershey, PA: IGI Global. doi:10.4018/978-1-5225-5216-1.ch009

Ramesh, M., Garg, R., & Subrahmanyam, G. V. (2017). Investigation of Influence of Quenching and Annealing on the Plane Fracture Toughness and Brittle to Ductile Transition Temperature of the Zinc Coated Structural Steel Materials. *International Journal of Surface Engineering and Interdisciplinary Materials Science, 5*(2), 33–42. doi:10.4018/IJSEIMS.2017070103

Robinson, J., & Beneroso, D. (2022). Project-Based Learning in Chemical Engineering: Curriculum and Assessment, Culture and Learning Spaces. In A. Alves & N. van Hattum-Janssen (Eds.), *Training Engineering Students for Modern Technological Advancement* (pp. 1–19). IGI Global. https://doi.org/10.4018/978-1-7998-8816-1.ch001

Rondon, B. (2021). Experimental Characterization of Admittance Meter With Crude Oil Emulsions. *International Journal of Electronics, Communications, and Measurement Engineering, 10*(2), 51–59. https://doi.org/10.4018/IJECME.2021070104

Rudolf, S., Biryuk, V. V., & Volov, V. (2018). Vortex Effect, Vortex Power: Technology of Vortex Power Engineering. In V. Kharchenko & P. Vasant (Eds.), *Handbook of Research on Renewable Energy and Electric Resources for Sustainable Rural Development* (pp. 500–533). Hershey, PA: IGI Global. doi:10.4018/978-1-5225-3867-7.ch021

Sah, A., Bhadula, S. J., Dumka, A., & Rawat, S. (2018). A Software Engineering Perspective for Development of Enterprise Applications. In A. Elçi (Ed.), *Handbook of Research on Contemporary Perspectives on Web-Based Systems* (pp. 1–23). Hershey, PA: IGI Global. doi:10.4018/978-1-5225-5384-7.ch001

Sahli, Y., Zitouni, B., & Hocine, B. M. (2021). Three-Dimensional Numerical Study of Overheating of Two Intermediate Temperature P-AS-SOFC Geometrical Configurations. In G. Badea, R. Felseghi, & I. Aşchilean (Eds.), *Hydrogen Fuel Cell Technology for Stationary Applications* (pp. 186–222). IGI Global. https://doi.org/10.4018/978-1-7998-4945-2.ch008

Sahoo, P., & Roy, S. (2017). Tribological Behavior of Electroless Ni-P, Ni-P-W and Ni-P-Cu Coatings: A Comparison. *International Journal of Surface Engineering and Interdisciplinary Materials Science, 5*(1), 1–15. doi:10.4018/IJSEIMS.2017010101

Sahoo, S. (2018). Laminated Composite Hypar Shells as Roofing Units: Static and Dynamic Behavior. In K. Kumar & J. Davim (Eds.), *Composites and Advanced Materials for Industrial Applications* (pp. 249–269). Hershey, PA: IGI Global. doi:10.4018/978-1-5225-5216-1.ch011

Sahu, H., & Hungyo, M. (2018). Introduction to SDN and NFV. In A. Dumka (Ed.), *Innovations in Software-Defined Networking and Network Functions Virtualization* (pp. 1–25). Hershey, PA: IGI Global. doi:10.4018/978-1-5225-3640-6.ch001

Salem, A. M., & Shmelova, T. (2018). Intelligent Expert Decision Support Systems: Methodologies, Applications, and Challenges. In T. Shmelova, Y. Sikirda, N. Rizun, A. Salem, & Y. Kovalyov (Eds.), *Socio-Technical Decision Support in Air Navigation Systems: Emerging Research and Opportunities* (pp. 215–242). Hershey, PA: IGI Global. doi:10.4018/978-1-5225-3108-1.ch007

Samal, M. (2017). FE Analysis and Experimental Investigation of Cracked and Un-Cracked Thin-Walled Tubular Components to Evaluate Mechanical and Fracture Properties. In P. Samui, S. Chakraborty, & D. Kim (Eds.), *Modeling and Simulation Techniques in Structural Engineering* (pp. 266–293). Hershey, PA: IGI Global. doi:10.4018/978-1-5225-0588-4.ch009

Samal, M., & Balakrishnan, K. (2017). Experiments on a Ring Tension Setup and FE Analysis to Evaluate Transverse Mechanical Properties of Tubular Components. In P. Samui, S. Chakraborty, & D. Kim (Eds.), *Modeling and Simulation Techniques in Structural Engineering* (pp. 91–115). Hershey, PA: IGI Global. doi:10.4018/978-1-5225-0588-4.ch004

Samarasinghe, D. A., & Wood, E. (2021). Innovative Digital Technologies. In J. Underwood & M. Shelbourn (Eds.), *Handbook of Research on Driving Transformational Change in the Digital Built Environment* (pp. 142–163). IGI Global. https://doi.org/10.4018/978-1-7998-6600-8.ch006

Sawant, S. (2018). Deep Learning and Biomedical Engineering. In U. Kose, G. Guraksin, & O. Deperlioglu (Eds.), *Nature-Inspired Intelligent Techniques for Solving Biomedical Engineering Problems* (pp. 283–296). Hershey, PA: IGI Global. doi:10.4018/978-1-5225-4769-3.ch014

Schulenberg, T. (2021). Energy Conversion Using the Supercritical Steam Cycle. In L. Chen (Ed.), *Handbook of Research on Advancements in Supercritical Fluids Applications for Sustainable Energy Systems* (pp. 659–681). IGI Global. doi:10.4018/978-1-7998-5796-9.ch018

Sezgin, H., & Berkalp, O. B. (2018). Textile-Reinforced Composites for the Automotive Industry. In K. Kumar & J. Davim (Eds.), *Composites and Advanced Materials for Industrial Applications* (pp. 129–156). Hershey, PA: IGI Global. doi:10.4018/978-1-5225-5216-1.ch007

Shaaban, A. A., & Shehata, O. M. (2021). Combining Response Surface Method and Metaheuristic Algorithms for Optimizing SPIF Process. *International Journal of Manufacturing, Materials, and Mechanical Engineering, 11*(4), 1–25. https://doi.org/10.4018/IJMMME.2021100101

Shafaati Shemami, M., & Sefid, M. (2022). Implementation and Demonstration of Electric Vehicle-to-Home (V2H) Application: A Case Study. In M. Alam, R. Pillai, & N. Murugesan (Eds.), *Developing Charging Infrastructure and Technologies for Electric Vehicles* (pp. 268–293). IGI Global. https://doi.org/10.4018/978-1-7998-6858-3.ch015

Shah, M. Z., Gazder, U., Bhatti, M. S., & Hussain, M. (2018). Comparative Performance Evaluation of Effects of Modifier in Asphaltic Concrete Mix. *International Journal of Strategic Engineering, 1*(2), 13–25. doi:10.4018/IJoSE.2018070102

Sharma, N., & Kumar, K. (2018). Fabrication of Porous NiTi Alloy Using Organic Binders. In K. Kumar & J. Davim (Eds.), *Composites and Advanced Materials for Industrial Applications* (pp. 38–62). Hershey, PA: IGI Global. doi:10.4018/978-1-5225-5216-1.ch003

Shivach, P., Nautiyal, L., & Ram, M. (2018). Applying Multi-Objective Optimization Algorithms to Mechanical Engineering. In M. Ram & J. Davim (Eds.), *Soft Computing Techniques and Applications in Mechanical Engineering* (pp. 287–301). Hershey, PA: IGI Global. doi:10.4018/978-1-5225-3035-0.ch014

Shmelova, T. (2018). Stochastic Methods for Estimation and Problem Solving in Engineering: Stochastic Methods of Decision Making in Aviation. In S. Kadry (Ed.), *Stochastic Methods for Estimation and Problem Solving in Engineering* (pp. 139–160). Hershey, PA: IGI Global. doi:10.4018/978-1-5225-5045-7.ch006

Siero González, L. R., & Romo Vázquez, A. (2017). Didactic Sequences Teaching Mathematics for Engineers With Focus on Differential Equations. In M. Ramírez-Montoya (Ed.), *Handbook of Research on Driving STEM Learning With Educational Technologies* (pp. 129–151). Hershey, PA: IGI Global. doi:10.4018/978-1-5225-2026-9.ch007

Sim, M. S., You, K. Y., Esa, F., & Chan, Y. L. (2021). Nanostructured Electromagnetic Metamaterials for Sensing Applications. In M. Bhat, I. Wani, & S. Ashraf (Eds.), *Applications of Nanomaterials in Agriculture, Food Science, and Medicine* (pp. 141–164). IGI Global. https://doi.org/10.4018/978-1-7998-5563-7.ch009

Singh, R., & Dutta, S. (2018). Visible Light Active Nanocomposites for Photocatalytic Applications. In K. Kumar & J. Davim (Eds.), *Composites and Advanced Materials for Industrial Applications* (pp. 270–296). Hershey, PA: IGI Global. doi:10.4018/978-1-5225-5216-1.ch012

Skripov, P. V., Yampol'skiy, A. D., & Rutin, S. B. (2021). High-Power Heat Transfer in Supercritical Fluids: Microscale Times and Sizes. In L. Chen (Ed.), *Handbook of Research on Advancements in Supercritical Fluids Applications for Sustainable Energy Systems* (pp. 424–450). IGI Global. https://doi.org/10.4018/978-1-7998-5796-9.ch012

Sözbilir, H., Özkaymak, Ç., Uzel, B., & Sümer, Ö. (2018). Criteria for Surface Rupture Microzonation of Active Faults for Earthquake Hazards in Urban Areas. In N. Ceryan (Ed.), *Handbook of Research on Trends and Digital Advances in Engineering Geology* (pp. 187–230). Hershey, PA: IGI Global. doi:10.4018/978-1-5225-2709-1.ch005

Stanciu, I. (2018). Stochastic Methods in Microsystems Engineering. In S. Kadry (Ed.), *Stochastic Methods for Estimation and Problem Solving in Engineering* (pp. 161–176). Hershey, PA: IGI Global. doi:10.4018/978-1-5225-5045-7.ch007

Strebkov, D., Nekrasov, A., Trubnikov, V., & Nekrasov, A. (2018). Single-Wire Resonant Electric Power Systems for Renewable-Based Electric Grid. In V. Kharchenko & P. Vasant (Eds.), *Handbook of Research on Renewable Energy and Electric Resources for Sustainable Rural Development* (pp. 449–474). Hershey, PA: IGI Global. doi:10.4018/978-1-5225-3867-7.ch019

Sukhyy, K., Belyanovskaya, E., & Sukhyy, M. (2021). *Basic Principles for Substantiation of Working Pair Choice.* IGI Global. doi:10.4018/978-1-7998-4432-7.ch002

Suri, M. S., & Kaliyaperumal, D. (2022). Extension of Aspiration Level Model for Optimal Planning of Fast Charging Stations. In A. Fekik & N. Benamrouche (Eds.), *Modeling and Control of Static Converters for Hybrid Storage Systems* (pp. 91–106). IGI Global. https://doi.org/10.4018/978-1-7998-7447-8.ch004

Tallet, E., Gledson, B., Rogage, K., Thompson, A., & Wiggett, D. (2021). Digitally-Enabled Design Management. In J. Underwood & M. Shelbourn (Eds.), *Handbook of Research on Driving Transformational Change in the Digital Built Environment* (pp. 63–89). IGI Global. https://doi.org/10.4018/978-1-7998-6600-8.ch003

Terki, A., & Boubertakh, H. (2021). A New Hybrid Binary-Real Coded Cuckoo Search and Tabu Search Algorithm for Solving the Unit-Commitment Problem. *International Journal of Energy Optimization and Engineering*, *10*(2), 104–119. https://doi.org/10.4018/IJEOE.2021040105

Tüdeş, Ş., Kumlu, K. B., & Ceryan, S. (2018). Integration Between Urban Planning and Natural Hazards For Resilient City. In N. Ceryan (Ed.), *Handbook of Research on Trends and Digital Advances in Engineering Geology* (pp. 591–630). Hershey, PA: IGI Global. doi:10.4018/978-1-5225-2709-1.ch017

Ulamis, K. (2018). Soil Liquefaction Assessment by Anisotropic Cyclic Triaxial Test. In N. Ceryan (Ed.), *Handbook of Research on Trends and Digital Advances in Engineering Geology* (pp. 631–664). Hershey, PA: IGI Global. doi:10.4018/978-1-5225-2709-1.ch018

Valente, M., & Milani, G. (2017). Seismic Assessment and Retrofitting of an Under-Designed RC Frame Through a Displacement-Based Approach. In V. Plevris, G. Kremmyda, & Y. Fahjan (Eds.), *Performance-Based Seismic Design of Concrete Structures and Infrastructures* (pp. 36–58). Hershey, PA: IGI Global. doi:10.4018/978-1-5225-2089-4.ch002

Vargas-Bernal, R. (2021). Advances in Electromagnetic Environmental Shielding for Aeronautics and Space Applications. In C. Nikolopoulos (Ed.), *Recent Trends on Electromagnetic Environmental Effects for Aeronautics and Space Applications* (pp. 80–96). IGI Global. https://doi.org/10.4018/978-1-7998-4879-0.ch003

Vasant, P. (2018). A General Medical Diagnosis System Formed by Artificial Neural Networks and Swarm Intelligence Techniques. In U. Kose, G. Guraksin, & O. Deperlioglu (Eds.), *Nature-Inspired Intelligent Techniques for Solving Biomedical Engineering Problems* (pp. 130–145). Hershey, PA: IGI Global. doi:10.4018/978-1-5225-4769-3.ch006

Verner, C. M., & Sarwar, D. (2021). Avoiding Project Failure and Achieving Project Success in NHS IT System Projects in the United Kingdom. *International Journal of Strategic Engineering*, *4*(1), 33–54. https://doi.org/10.4018/IJoSE.2021010103

Verrollot, J., Tolonen, A., Harkonen, J., & Haapasalo, H. J. (2018). Challenges and Enablers for Rapid Product Development. *International Journal of Applied Industrial Engineering*, *5*(1), 25–49. doi:10.4018/IJAIE.2018010102

Wan, A. C., Zulu, S. L., & Khosrow-Shahi, F. (2021). Industry Views on BIM for Site Safety in Hong Kong. In J. Underwood & M. Shelbourn (Eds.), *Handbook of Research on Driving Transformational Change in the Digital Built Environment* (pp. 120–140). IGI Global. https://doi.org/10.4018/978-1-7998-6600-8.ch005

Yardimci, A. G., & Karpuz, C. (2018). Fuzzy Rock Mass Rating: Soft-Computing-Aided Preliminary Stability Analysis of Weak Rock Slopes. In N. Ceryan (Ed.), *Handbook of Research on Trends and Digital Advances in Engineering Geology* (pp. 97–131). Hershey, PA: IGI Global. doi:10.4018/978-1-5225-2709-1.ch003

You, K. Y. (2021). Development Electronic Design Automation for RF/Microwave Antenna Using MATLAB GUI. In A. Nwajana & I. Ihianle (Eds.), *Handbook of Research on 5G Networks and Advancements in Computing, Electronics, and Electrical Engineering* (pp. 70–148). IGI Global. https://doi.org/10.4018/978-1-7998-6992-4.ch004

Yousefi, Y., Gratton, P., & Sarwar, D. (2021). Investigating the Opportunities to Improve the Thermal Performance of a Case Study Building in London. *International Journal of Strategic Engineering*, *4*(1), 1–18. https://doi.org/10.4018/IJoSE.2021010101

Zindani, D., & Kumar, K. (2018). Industrial Applications of Polymer Composite Materials. In K. Kumar & J. Davim (Eds.), *Composites and Advanced Materials for Industrial Applications* (pp. 1–15). Hershey, PA: IGI Global. doi:10.4018/978-1-5225-5216-1.ch001

Zindani, D., Maity, S. R., & Bhowmik, S. (2018). A Decision-Making Approach for Material Selection of Polymeric Composite Bumper Beam. In K. Kumar & J. Davim (Eds.), *Composites and Advanced Materials for Industrial Applications* (pp. 112–128). Hershey, PA: IGI Global. doi:10.4018/978-1-5225-5216-1.ch006

About the Contributors

Ashok Sharma has 18 Years of Teaching Experiences in Higher Education and i have worked in Various Reputed Institution of Higher Learning in India in different capacity.My area of Interest is Machine Learning, Cloud Computing and Data Science. He has attended 40+ DST/AICTE sponsored Training in different Technologies and have 40+ Research Articles in SCI/SCIE/Scopus indexed Journals with 10 Patent and Four Copyright, has developed MooCs in 4 Quadrant of 15 Courses, Running online Courses on Virtual Platform like Moodle,MoodleCloud,Canvas and Blackboard LMS .5 PhD has been awarded and 4 Scholars are working under my guidance in the area of Cloud Computing, Data Science and Cognitive Behaviour Analysis.

Sandeep Singh Sengar is a Lecturer in Computer Science at Cardiff Metropolitan University, United Kingdom. He also holds the position of Cluster Leader for Computer Vision/Image Processing at this place. Before joining this position, he worked as a Postdoctoral Research Fellow at the Machine Learning Section of the Computer Science Department, at the University of Copenhagen, Denmark (a ranked #1 university in Denmark). He completed his Ph.D. degree in Computer Vision at the Department of Computer Science and Engineering from the Indian Institute of Technology (ISM), Dhanbad, India, and an M. Tech. degree from Motilal Nehru National Institute of Technology, Allahabad, India. He is also a Fellow of HEA, a Senior Member of IEEE, and a Professional Member of ACM. Dr. Sengar's broader research interests include Machine/Deep Learning, Computer Vision, Image/Video Processing, and its applications. He has published several research articles in reputable international journals and conferences. He is an Editorial Board Member, Guest Editor, and Reviewer at reputed International Journals. He is a reviewer of research grants at EPSRC and Cardiff Met.

Parveen Singh is an Associate Professor and Head Department of Computer Sciences,Govt SPMR College of Commerce, Jammu, honoured with National Award by Govt of India, for popularization of Science among children's, is a dynamic instructor and thought leader, focused on providing students with a rigorous and challenging education. Earned recognition as a knowledgeable teacher, author along with as a debater with well-organized, stimulating, and student-centred courses. Cultivated teaching partnerships and alliances with key business contacts across the J&k . Co-authored 09 Books and attended more than 17 international and national conferences. Certified, Research Based Pedagogical tools Trainer from Centre Of Excellence in Science and Mathematics Education IISER pune. Awarded with Govt. of J&K, Science Innovative Teacher Award 2013, Awarded with Govt of Tamil Nadu ICTACT Best Techno Faculty Award 2016, Awarded with Principal Citation for contribution towards Growth of Institution 2010,Awarded with Distinguished HOD award 2017, by Computer Society of India, CSI, Mumbai Chapter, Member of Jammu University Academic council from 2006-2009,Member of Board of Studies, Jammu University since 2003.

B. Shamreen Ahamed holds a Bachelor's degree in Computer Science and Engineering from KCG College of Technology, Chennai. She is a Gold Medalist and a University Rank holder for her Master's degree in Computer Science and Engineering from Sathyabama University. Shamreen is currently doing Ph.D. in Computer Science and Engineering with major in Machine Learning. Her specializations include Machine Learning, Deep Learning and Artificial Intelligence.

Subhabrata Barman is an Assistant Professor with the Department of Computer Science & Engineering, Haldia Institute of Technology, West Bengal, India. His research interests are in the field of Wireless Networks, Computational Intelligence, Remote Sensing and Geo-Informatics, Parallel and Grid Computing. He has published research papers at various International and National Journals and Conferences. He is a Professional Member of IEEE, IACSIT, IAENG and a reviewer of International Journal of Wireless Networks (Springer).

Nikita Chauhan is a final year student pursuing Bachelors in Technology in the field of Information Technology.

Arjun Choudhary received his B.Tech degree in Communication and Computer Engineering from The LNMIIT Jaipur, India, in 2009. He did his post-graduation (M.Tech.) in Computer Science and Engineering from the Motilal Nehru National Institute of Technology Allahabad, India, in 2012. He received his Ph.D. degree in Computer Science and Engineering, JNVU, Jodhpur in 2022. His research topic of doctoral degree is in Cloud computing and Security. He currently works as an Assistant Professor at the Department of Computer Science and Engineering and heading the Centre for Cyber Security, SPUP, Jodhpur, India as Deputy Director. His current research interests include Cloud Computing, Information Security, Digital forensics, and Deep Learning.

Asha Karegowda is currently working as Associate Professor in Dept of MCA, Siddaganga Institute of Technology, Tumkur, Karnataka, India, since 1998. Her area of interest include Data mining, WSN, Remote sensing, Bio inspired computing and Deep learning.She has authored few books on C, Data structures using C, Python and Data mining. She has published few book chapters; and papers in conferences and journals. She is currently Guiding few research scholars in the area of image processing and remote sensing. She is handling subjects for both PG and UG students: Python, Data structures using C and C++, Data mining, Data Analytics, Big data, Digital marketing etc.

Santanu Koley earned his doctorate in philosophy (PhD) from CSJM University in Kanpur, Uttar Pradesh, India in 2013, and he is currently employed as a professor in the department of computer science and engineering at Haldia Institute of Technology in Haldia, West Bengal, India. In addition to sixteen years of teaching experience, he has more than fourteen years of research experience from several AICTE-approved engineering colleges across India. Dr. Koley has published more than 30 research papers in journals and conferences from throughout the nation and the world. The areas of cloud computing, digital image processing, artificial intelligence, and machine learning are where he is currently concentrating his research efforts.

Meenakshi S. Arya is an experienced academician with a demonstrated history of working in the education management industry. Skilled in Machine Learning, Big Data Analytics, Algorithms, Data Structures, and C, C++ (Programming Language). Strong education professional with a Ph.D focused in Computer Science and Engineering from Jaypee University of Information Technology. Currently She is working as Professor and Associate Director in School of Artificial Intelligence, MIT(World Peace University).

N. Sasikaladevi received Ph. D. Degree in Computer Science in 2013. She has published more than 42 papers in reputed International journals and conferences including publications in SCI-Indexed Journals. She is a reviewer of more than a dozen of reputed journals including IEEE transactions on Services Computing and IEEE Journal of Internet of Things. s. Her research focus includes the design of machine learning strategies to solve Discrete Logarithm Problem. She published several books and chapters in reputed publisher including Prentice Hall of India, Lambert Academic Publisher, IGI Global, Springer and Science Direct. She has received fund from Department of Science and Technology, Government of India to carry out projects in the domain of Services computing and Security System. She has received young scientist award and women scientist award from DST, India.

She is also involved in the security enriched payment system design as a part of Digital India Initiative. Her current research interests include information security, digital authentication, security of wireless sensor networks, computer vision, deep learning and security of vehicular ad-hoc networks.

P. Prabhavathy - Contributing Author| **Prabhavathy Pachaiyappan** received the Bachelor's degree in Computer Science in 2005, the M.E, in 2010, and the Ph.D. degree from Anna University, in 2017. From 2011 to 2020, she worked as Teaching Fellow at Anna University. She is currently an Assistant Professor (Selection grade) in the Department of Computer Science and Engineering at SRM Institute of Science and Technology. Her research interests are in Networking

S. Hemalatha received her BE in Computer Science and Engineering from the University of Madras, TN, India in 2000 and M.Tech in Computer Science and Engineering in 2004. She submitted her Ph.D thesis in the Image Processing domain. She is working as an Assistant Professor (Selection Grade) in VIT University, TN, India. She has about 17 years of teaching experience in the field of Computer Science and Engineering. Her research interests include Image Processing, Pattern Recognition, Image Classification and Segmentation. She published papers with International journals and international conferences in these areas.

Athira P. Shaji received her BTech in Computer Science and Engineering from the University of Kerala, India in 2015 and M.Tech in Computer Science and Engineering in 2018 from Mahatma Gandhi University, Kerala. She is pursuing her PhD in Image Processing and Machine Learning technologies in VIT University, Vellore. She has about 7 years of teaching experience in the field of Computer Science and Engineering. Her research interests include Image Processing, Computer Vision, Machine Learning and Deep Learning.

Ajay Sharma is pursuing a PhD from Manit and Teaching Fellow at VIT Bhopal University.

Bhavana P. Shrivastava is currently working as Assistant Professor In MANIT,Bhopal.She did her B.E (Electronics and Telecommunication Engineering), M. Tech (Digital Communication) and Ph.D. (VLSI) from Maulana Azad National Institute of Technology, Bhopal, Madhya Pradesh. She has Completed many projects of Image Processing and VLSI Designs and having 17 years of teaching experience. Publish many papers in reputed journals and conferences. She is Senior member of IEEE, lifetime member of ISTE and IETE.

Rajnesh Singh currently works at the GL Bajaj Institute of Technology and Management, Greater Noida, India. Rajnesh pursed his Master of Technology in Computer Science and Engineering from CDAC Noida, Affiliated to Guru Gobind Singh Indraprastha University, New Delhi and Ph.D. from Gautam Buddha University, Greater Noida in the area of Mobile Adhoc Networks. He is a seasoned academician having more than 16+ years of experience. He has more than 30+ publications in reputed International Journals and Conferences along with some paper presentations in National Conferences. His one paper is judged for best paper award in IEEE Conference. His current areas of research are networks, mobile Adhoc network, algorithms, IoT and Machine Learning etc. He is also a member of ACM Society.

Ashish Tripathi received the M.Tech degree in Computer Science and Engineering from the Motilal Nehru National Institute of Technology Allahabad, India, in 2012 and a Ph.D. degree in Computer Science and Engineering from Motilal Nehru National Institute of Technology Allahabad, in 2015. He currently works as an Associate Professor at the Department of Information Technology, G. L. Bajaj Institute of Technology and Management, Greater Noida, India. His current research interests include Optimization Techniques, Soft Computing, Evolutionary Computation, Machine Learning, Deep Learning, and Cloud Computing.

Auxilia Osvin Nancy Vincent obtained her Bachelor's degree in Information Technology from Madras University. Then she obtained her Master's degree in Information Technology from Sathyabama University and currently doing PhD in Computer Science majoring in Deep learning. Her specializations include Deep learning, Machine learning, and Digital Video and Image Processing.

Index

A

Anatomy 77, 101, 103, 110-111, 114, 116
Applications 1, 6, 9-10, 12, 28, 32, 34, 36,
        43, 52-54, 62-63, 67, 70, 73, 98-100,
        103, 111, 113, 117-118, 128, 133,
        136, 153-155, 161, 164, 166, 171-172,
        180, 182-183, 185, 190, 193-194, 196
AR healthcare 103
Artificial Intelligence 2, 10, 64, 100-101, 110,
        154-155, 164, 171, 181-182, 186-197
Artificial Neural Network 155, 167
Augmented Reality 76, 101, 103-117

C

CNN 13, 20, 75, 78, 98, 119-121, 126,
        134-138, 140-141, 143-145, 147-148,
        150-151, 155-157, 162, 171, 173-174,
        176, 182, 184
Computer vision 6, 12-13, 29-30, 34, 38,
        64-65, 67, 70, 72-73, 84, 94, 136, 151,
        169-175, 181, 183-184
Convolutional Neural Network 30, 75, 119,
        121-122, 126, 132, 143, 151-153, 155,
        166-167, 171, 174, 184
Customer Satisfaction 39, 186, 191, 193

D

Decision Making 187-188, 190-192
Deep Learning 12-13, 29-30, 75-76, 97,
        99, 101, 119-121, 123, 126, 128-134,
        136-137, 147-148, 151, 153-155, 161,
        163-166, 168-169, 171, 174-176, 180-
        181, 184, 194

DenseNet-201 119, 122-123
dermoscopic images 134, 138, 140
Down-sampling 12

E

edge detection 65, 69-70, 84-87, 89, 91, 94,
        97-98, 100, 102, 170, 182
edge detection operators 65, 70, 89, 91
evaluation 26-28, 32, 34, 44, 46, 49, 51-52,
        54, 56-57, 60, 62-63, 65, 98, 116, 121,
        130, 135, 138, 140, 142, 154, 159-162,
        164-165, 174, 176
Evolutionary 1, 3, 8-10, 34, 197
Extended Reality 104-105, 110
Eye Disease 119-120, 133

F

Feature extraction 24, 68, 83, 136, 144-145,
        153-156, 159, 161, 163, 169-170, 172,
        174, 185

G

Gradient-Based 1, 3-4, 7-8, 11, 33-34, 52,
        54, 61

H

hyper meta learning 32, 34, 51-54, 57-59, 62-64

I

image gradient 84
Image processing 10, 31, 65, 67-68, 70-71,

75, 83-84, 88, 94, 98-102, 123, 126, 129, 133, 169-170, 172, 175, 184
image segmentation 11, 69-71, 73-75, 77, 84, 98, 100-102, 154, 159, 161, 165, 167, 174
Image Super Resolution 12, 14, 27
ISIC 2019 140, 148

L

Laplacian of Gaussian (loG) 87, 89, 102

M

Machine learning 1-3, 9-11, 32-37, 40-42, 44, 46-51, 53-64, 67, 73, 83, 101, 123, 132-133, 136, 151, 154, 169-172, 175, 178, 181-184, 187-188, 194-196
MAML 1, 4-7, 9, 11, 34
Management 80, 103, 163, 169, 186-187, 189, 191-192, 194-196
medical surgeries 103
Memory-Augmented 3, 7-8, 11
Meta Learning 2, 10-11, 32, 34, 37, 51-54, 57-59, 62-64
MobileNet 135, 140-143, 147, 149-151

O

Optimization 1-4, 9-11, 33-34, 39-52, 54, 56-64, 126-128, 174-175, 181, 183, 185, 191

P

Perceptual Quality 13
Plant disease recognition 169, 173, 183-184

pretrained model 136-138, 140

R

Recurrent Neural Network 8, 155, 168
Remote-sensing Imagery 165
Retinal Fundus Images 119-121, 131-133

S

Segmentation 11, 67-80, 82-84, 89, 97-102, 123, 125, 133, 136, 153-161, 163-168, 170, 172-175, 178, 181, 185
selection 8, 11, 32-33, 42, 44-47, 49-51, 54-55, 63, 126, 133, 157, 167, 170-171, 174-175, 183, 191
Skin lesion 135-137, 140-142, 145, 150, 152

T

Transfer Learning 9, 33, 36-37, 59, 62, 123-125, 132, 151-152, 174, 176

U

Up-Sampling 24

V

VGG 16 136
VGG19 135, 140-141

W

Water-bodies 154-155, 163, 167

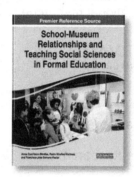
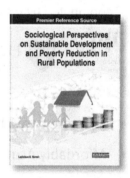

Printed in the United States
by Baker & Taylor Publisher Services